IIASA PROCEEDINGS SERIES

Volume 4

Pest Management

IIASA PROCEEDINGS SERIES

VOLUME 1 — CARBON DIOXIDE, CLIMATE AND SOCIETY
Proceedings of a IIASA Workshop Cosponsored by WMO, UNEP, and SCOPE, February 21-24, 1978
Jill Williams, Editor

VOLUME 2 — SARUM AND MRI: DESCRIPTION AND COMPARISON OF A WORLD MODEL AND A NATIONAL MODEL
Proceedings of the Fourth IIASA Symposium on Global Modelling, September 20-23, 1976
Gerhart Bruckmann, Editor

VOLUME 3 — NONSMOOTH OPTIMIZATION
Proceedings of a IIASA Workshop,
March 28-April 8, 1977
Claude Lemarechal and Robert Mifflin, Editors

VOLUME 4 — PEST MANAGEMENT
Proceedings of an International Conference,
October 25-29, 1976
G.A. Norton and C.S. Holling, Editors

FORTHCOMING:

METHODS AND MODELS FOR ASSESSING ENERGY RESOURCES
First IIASA Conference on Energy Resources,
May 20-21, 1975
Michel Grenon, Editor

FUTURE COAL SUPPLY FOR THE WORLD ENERGY BALANCE
Third IIASA Conference on Energy Resources,
November 28-December 2, 1977
Michel Grenon, Editor

THE SHINKANSEN HIGH-SPEED RAIL NETWORK OF JAPAN
Proceedings of a IIASA Conference,
June 27-30, 1977
A. Straszak and R. Tuch, Editors

PEST MANAGEMENT

Proceedings of an International Conference,
October 25-29, 1976

G.A. NORTON
Imperial College Field Station
Ascot, Berkshire
United Kingdom

C.S. HOLLING
International Institute for Applied Systems Analysis
Laxenburg, Austria

Editors

PERGAMON PRESS

OXFORD · NEW YORK · TORONTO · SYDNEY · PARIS · FRANKFURT

U.K.	Pergamon Press Ltd., Headington Hill Hall, Oxford OX3 0BW, England
U.S.A.	Pergamon Press Inc., Maxwell House, Fairview Park, Elmsford, New York 10523, U.S.A.
CANADA	Pergamon of Canada, Suite 104, 150 Consumers Road, Willowdale, Ontario M2 J1P9, Canada
AUSTRALIA	Pergamon Press (Aust.) Pty. Ltd., P.O. Box 544, Potts Point, N.S.W. 2011, Australia
FRANCE	Pergamon Press SARL, 24 rue des Ecoles, 75240 Paris, Cedex 05, France
FEDERAL REPUBLIC OF GERMANY	Pergamon Press GmbH, 6242 Kronberg-Taunus, Pfedrstrasse 1, Federal Republic of Germany

First edition 1979

British Library Cataloguing in Publication Data

Pest Management. - (International Institute for Applied Systems Analysis. IIASA Proceedings Series; vol. 4).
1. Pest Control - Mathematical models - Congresses
2. Systems analysis - Congresses
I. Norton, G A II. Holling, C S III. Series
632'.9'02851 SB950.A2 78-40825

ISBN 0 08 023427 5

Acknowledgement

Ecological Policy Design: A Case Study of Forest and Pest Management by C. S. Holling, Dixon D. Jones and William C. Clark is reproduced by permission of John Wiley and Sons, Limited, from *Adaptive Environmental Assessment and Management*, edited by C. S. Holling, published by John Wiley and Sons, Ltd., Baffins Lane, Chichester, 1978.

Printed and bound at William Clowes & Sons Limited Beccles and London

PREFACE

During the past five years, progress has been made in applying mathematics and policy analysis to ecological problems, and an increasing effort has been made to apply systems analysis techniques to pest management problems. Developments in this field have taken place worldwide, including at the International Institute for Applied Systems Analysis (IIASA), where the Ecology/Environment Program has concentrated on the management problems associated with spruce budworm in North American forests, using it as a prototype.

As it was felt that this research had reached the stage at which workers could usefully exchange experiences, a Pest Management Conference was held by IIASA in Laxenburg, Austria, October 25-29, 1976. Thirty-one people, representing 10 National Member Organizations (see Appendix 1), attended the conference.

During the first three days of the conference, papers were presented by the IIASA/University of British Columbia group and the Environmental Management Unit of Imperial College, London, and by other workers from IIASA, France, Japan, the USA, and the USSR. The fourth day was devoted to working sessions, at which participants discussed where developments in modelling techniques should be encouraged and how their use in pest management might be improved. Finally, recommendations were made concerning the role that IIASA might play in the future in applying systems analysis to pest management.

CONTENTS

RECOMMENDATIONS AND REPORTS OF WORKING GROUPS

RECOMMENDATIONS

To stimulate discussion within IIASA concerning its future role in pest management, the following recommendations were made:

1. An IIASA Team

Pest management should be continued as a major activity within IIASA. It should concentrate on the development and fostering of case studies and particularly on the role of implementation. The feasibility of this option clearly depends upon the presence of a small core of expertise. Such a core could be comprised of a revolving set of individuals drawn from existing projects in various countries. The emphasis would be upon distilling general lessons and encouraging the application of tested methodologies and concepts.

2. An IIASA Pest Management Network (PMN)

It is suggested that IIASA should act as the central focus of an institutional network, concentrating on the exchange of information between various pest management modelling groups. By performing this function, IIASA could remove some of the difficulties involved in communication between ongoing modelling programs. It is recommended that:

(i) IIASA sponsor a biennial conference/workshop on pest management to be held at Schloss Laxenburg.

(ii) IIASA act as a clearing house for information on pest management modelling. First, a list of workers actively engaged in the field should be compiled; a start has already been made (see Appendix 2). As preprints of these workers become available, they should be sent to IIASA where they would be duplicated and disseminated as working papers in the IIASA Pest Management Network Series.

3. An IIASA Focus on Methodological Problems in Pest Management Modelling

From discussions in the working groups several areas were identified where it was thought that methodological development, particularly within IIASA, should be encouraged. It is suggested that the following areas should be given priority:

(a) qualitative modelling techniques
(b) control and modelling of stochastic systems
(c) mathematical techniques of compression
(d) robust identification methods
(e) methods for dealing with spatially distributed systems.

REPORTS OF WORKING GROUPS

During the working session on the fourth day, participants divided into two discussion groups. The first discussion group (reporter - Dr. Don DeMichele) concentrated on the descriptive role of modelling and its use as a research tool. The second group (reporter - Dr. Christine Shoemaker) discussed the prescriptive role of modelling and its value in pest management. A summary of these discussions is given below.

Report of Working Group 1

The use of mathematical modelling techniques in the natural sciences is a relatively recent development and yet several distinct phases in its evolution can already be identified, depending on the dominant school of thought in the modelling community at the time. Initially, the regression modelling school held the stage. Here, large data sets were used to recall past observations as a means of predicting expected responses in the future. The subsequent school was concerned with attempts to construct giant, one-to-one models, involving contributions from various collaborators to detailed components of the overall model. This, in turn, was followed by a simplified modelling approach. In this case, and in contrast to the previous school, a simple notion is used as the basis of the model, additional components being added until an acceptable model response is achieved.

Upon evaluation, however, each approach, in some sense, has been found to be lacking, and it now appears that the modelling community in the natural sciences is beginning to accept the possibility that there is no one best modelling method. From the diversity of approach and range of modelling objectives found in the papers presented at this meeting, it is clear that the objectives of the modeller, the level of detail considered and the structure of the system itself have dictated the approach.

In attempting to model pest management problems, the purpose of modelling is to develop a coherent and logical procedure to deal with a complex system, described by incomplete knowledge and data. Faced with this situation, the modeller attempts to reconcile noisy data and hazy concepts within a consistent and robust model. His task is made more difficult with the ever increasing variety of techniques being introduced into pest management modelling.

Nevertheless, without new techniques, progress in pest management modelling would decline. Consequently, several research areas have been identified where the development of better modelling techniques is thought to be both feasible and desirable:

1. Competition

Competition occurs at a specific trophic level and can be inter- or intra-specific in nature. The problem is confounded by age structure considerations, by trophic switching and by feeding preferences. Although current mathematical formulations encapsulate certain aspects, they are unable to compress or fully integrate these biological processes.

2. Structural Stability

Discussions also indicated that the general area of ecosystem structural stability was one which deserved serious analytical study. The need for such studies arises from the fact that ecosystems are manipulated to control pest populations. By changing ecosystem parameters, however, radical changes in other populations within the ecosystem may occur. Hence, it is important that the structural stability of these systems, in the face of management perturbations, be investigated.

3. Qualitative Factors

The qualitative differences that occur between individuals in plant and animal populations can greatly influence their population dynamics; a classic example of the importance of qualitative differences in a pest population being found in the migratory locust. Qualitative differences can occur within individuals as they age, and within populations over time. However, even though the physiological basis for many of the processes involved are well investigated (e.g. morph determination in aphids), no conceptual framework for handling this class of problem exists at present. More effort should be devoted to the measurement of these qualitative changes in populations and to the development of a concise, yet robust, mathematical description that can be incorporated in the general framework of population dynamics.

4. Communication Techniques

As demonstrated in these proceedings, the level of complexity of pest management models varies considerably, depending in part on the modeller's experience and mathematical expertise. In most cases, the modeller attempts to reduce the complexity of a model, while retaining the essence of the problem. In attempting to compress information contained in and derived from complex models, we feel there is considerable room for developing appropriate descriptive techniques--such as graphical techniques, manifold analysis and parametric analysis. This would not only improve communication within the modelling community but also between the modeller and the pest manager.

5. Component Library

It is suggested that the modelling of ecological as well as individual animal and plant systems has reached such a stage that it would be feasible to outline generally accepted notions and procedures. This component library, in the form of a textbook or extensive review paper, would be of significant help in developing standard lines of communication between workers in the field.

Conclusion

As the field of pest management modelling continues to develop, attention will focus on the techniques of modelling and their use. To handle populations with diverse and time dependent characteristics, more powerful analytical techniques need to be developed. Analysis methods must also be developed for studying the stability properties of the ecosystems involved. In the future, emphasis must be placed on the general nature of modelling effort rather than the specifics of a particular case. Where possible, universal components and notions should be utilized to reduce the degree of complexity for the rest of the modelling community and their clients. Of equal importance is the need to develop general methods of analyzing the accuracy and robustness of pest management models. Agreement with data does not of itself validate a model, data can only disprove the notions of the model.

Report of Working Group 2

To investigate the prescriptive role of pest management modelling, discussion was directed towards three fundamental questions: (1) why build a model? (2) which factors determine its structure and complexity? and (3) what impact has pest management modelling had in the past?

Why construct a model to deal with a pest management problem?

Modelling in a pest management context may not only improve our understanding of the biology of the system involved but may also assist in the prescription of improved pest management strategies. In a research role, models can serve as a means of organizing currently available data as well as indicating those areas which require further studies. In the area of prescription, models can help to determine more efficient ways of using currently available technology (e.g. improved timing of insecticide treatment), in predicting the usefulness of possible new technology (e.g. resistant plant varieties), and in examining the effects of government intervention on pest management (e.g. taxes on environmentally hazardous insecticides or land use controls).

At this conference, the areas that have been most thoroughly explored are those of scientific description and the improved use of existing technology. For instance, the papers by Gutierrez and Kiritani have illustrated the way in which models can increase biological understanding. Those by Norton and Shoemaker show how models can be used to improve the application of existing control measures, such as insecticide application and time of harvesting. Wider policy issues associated with pest management are discussed by Holling et al.

Which factors determine the structure and complexity of a pest management model?

With available resources and data, the system analyst's task is to determine which of a number of possible objectives the model can realistically hope to meet. Thus, the objectives of modelling will depend upon:

(a) the data currently available, that which can be collected during the study, and data that is available to the grower, such as a forecast or monitored information,

(b) the feasible management alternatives,

(c) the effect of these alternatives on biological interactions.

The structure and complexity of the model, and the extent to which it provides a quantitative description of the effect of pest management variables on agricultural or forest ecosystems depends upon these three variables, but particularly the data available. For instance, detailed information on development rates and age specific mortality, both from natural causes and from insecticide treatment, is necessary for constructing a model to determine the best timing of insecticide treatments. If the model is to be of the economic threshold type, then monitored or forecast information must also be available.

What impact on actual pest control practice has modelling effort had in the past?

Without detailed information on the value of past modelling efforts in pest management, it was felt that a rigorous analysis of pest management models should be undertaken to assess the factors that determine successful implementation or failure. It would be hoped that such an analysis would be of value in the future in allocating resources between data collection, model construction and implementation.

The types of questions that might be asked in setting suitable guidelines for current and future projects would include:

(i) Does the study promote a demonstrably superior redistribution of resources in terms of research priorities?

(ii) Does the study allow a more effective use of available information in providing a critical understanding of a complex situation?

(iii) Does the study lead the decision maker to an appreciation of a wider range of management alternatives?

(iv) Have decision makers expressed a confidence in the approach, as reflected through a commitment of time and resources?

THE INSTITUTE OF RESOURCE ECOLOGY; UNIVERSITY OF BRITISH COLUMBIA, VANCOUVER, B.C., CANADA

INTRODUCTION

As the very broad scope of environmental and resource problems became clear in the mid-1960s, two trends began to emerge. First, a number of countries began to formalize an ecological policy design process whereby attempts were made to predict the impacts of specific proposed developments. These policy designs were derived largely from data surveys and the use of crude predictive techniques. The second trend that emerged in parallel to this one was the formation of large interdisciplinary teams to address resource management problems by drawing upon a diverse range of information and expertise from a number of disciplines. Some of these teams began to use the techniques emerging from computer applications to develop models they hoped would incorporate reasonable representations of ecological processes and produce realistic behavior. These were nobly motivated efforts, often expensive and experimental because they were so new. It is that history of experience, of success and of failures, that has led to tested concepts and techniques that deserve broader application. The failures were both expected and necessary; that is how we learn. These failures were the consequences of inexperience in ways to bridge the gaps between disciplines, data, techniques, knowledge, institutions, and people.

The Early Stages of the Institute of Resource Ecology (IRE) Program

We formally began to explore ways to bridge these gaps in 1967, concentrating on a sequence of steps, starting with disciplinary gaps (the easiest to bridge) and continuing to institutional gaps (the most difficult). Throughout, the steps which were most useful were the ones that were biased toward being practical, of short duration, involving a small core of people, and modest in their scope, resources, and objectives. But as a consequence, we found each step covered progressively more ground leading ultimately to a framework for a new science of ecological policy design. That bias colors much of what follows.

The first step was to apply and test techniques to bridge gaps between disciplines and between data sources. The IRE addressed particular, well defined resource management problems by drawing upon available expertise in a series of intense workshops. These early meetings led to techniques to integrate and focus the research of subject matter specialists by translating their various conceptual models and data into computer models. The resulting models, however, were less important in themselves than as a useful focus for discussion and for data organization.

The second step expanded the effort to begin bridging not just disciplines but some of the roles played in different institutions—between scientist, manager, and decision maker. That led to ways not only to describe more confidently the ecological and social dimensions of a problem but to define as well alternative management objectives, actions, and policies. As techniques were modified and tested, the subsequent workshops evolved from ones that required weekly meetings for many months to five-day exercises, which in themselves produced a product that could be viewed as a very preliminary but broad evaluation of a proposed policy or development. The workshops covered a variety of topics—recreational development, fisheries and wildlife policies, watershed and forest management, pest management, regional resource development, and hydroelectric power development.

That set of experiences established an approach to simulation modelling that emphasized (1) management questions, (2) a concise description of the dynamics of the biological and economic parts of the problem, and (3) the comparison of time streams of indicators generated by alternative policies. All these activities could be done by a core group of three or four analysts, in a period of one week, drawing upon the expertise of up to twenty specialists—scientists, managers, and decision makers. Most examples involved government agencies, but a few included industrial groups (e.g., hydroelectric development in Canada's north) and one included a group of citizens (high alpine development in Austria). But although the techniques were developed in this environment, they could not be viewed as tested or as useful for anything other than to focus dialogue.

That led to the third step—to undertake two thorough, long-term case studies to test rigorously the existing methods and to add others where necessary. And testing meant testing both theoretical rigour and practical usefulness. The best approach is hardly worthwhile unless someone can use it. And the most usable technique is dangerous if it lacks substance. Hence the ultimate goal of each case study was both a rigorous testing and analysis of methods and, as well, formal implementation within operating management agencies. Two case studies were chosen as prototypical examples—one of a forest insect pest system and one of a multi-species fisheries management and enhancement problem. The spruce budworm/forest problem in eastern North America has gone the furthest in coping with existing management questions, validating alternative modelling techniques, generating management alternatives, and evaluating their consequences. It has resulted in agencies of two Canadian provinces adopting the approach for the setting of research priorities and the development and evaluation of management options. The Pacific salmon case study has gone furthest in affecting and modifying a proposed new development to enhance fisheries populations.

This testing has led to the framework of an ecological policy design process of which impact assessment is only a part. But in order to subject that approach to a broader evaluation a fourth step was taken.

The Present Stage of the IRE

That fourth step formed what we unimaginatively called Group X, a group whose purpose was to test and evaluate the techniques and concepts mentioned above, adding others where necessary and feasible. It drew upon a network of expertise developed at IIASA and extended the experience of the Canadian group of the IRE and Environment Canada, to include experience with renewable resource problems in different national settings: renewable resource management and disease control in Venezuela and Argentina; range and wildlife management in the USA; oceanographic problems in Europe; ecological process studies in the USSR.

Any small group is biased, and indeed the individuals were chosen, ideally, for a defined and focussed bias. In this case the biases took the following form:

(1) No one was an expert in institutional organization and design.

(2) There was a bias towards experience and competence in mathematics, ecology, and in dealing with government management agencies--in short a simultaneous emphasis on rigour of analysis, relevance of concepts, and usefulness of technique.

(3) There was a bias that both theoretical and applied techniques had gone far beyond the state-of-the-art as it is practiced in ecological policy design. There are tested and practical quantitative techniques to model essential elements of ecological problems, and these are not expensive in resources or expertise. These techniques can be combined with prescriptive techniques for policy design and evaluation that together result in a framework for an adaptive management process. Finally, in addition to these tested techniques, there seemed to be a great potential for the development of qualitative, but systematic, techniques that could certainly be used to compress understanding of complex systems and perhaps even be used to model poorly known systems.

(4) There was a bias that the process and the product are inextricably linked; the sequence and design of the

workshops, the emphasis on adaptive approaches, and the design of different modes of communication are as important as the models and the analyses.

(5) Finally, understanding the way systems respond to disturbance is an essential step in identifying, classifying, and living with the unexpected.

These biases matched the purpose of our present program. First, we wished to synthesize and generalize our experience and knowledge of ecological systems as they relate to policy design. This has led to some conceptualizations that emphasize the key structure and behaviour of ecological systems underlying any effort to manage them.

The second goal was to evaluate experimentally the usefulness of a number of modelling and mathematical techniques that have been used or potentially could be used in ecological policy design. These experiments used groups of individuals to test the predictive power of a range of techniques selected to cover a spectrum from highly qualitative to highly quantitative. The strengths and weaknesses of each technique were tested using several specific case studies and for different amounts of information.

Third, we wished to summarize and evaluate our experience with both the process and methods of policy design. We found that to define and focus a problem, the workshop process, involving scientists, managers, and policy people, worked very well. During such a workshop, impact categories and objectives are classified, key information needs defined, alternative actions described, and the framework and crude working version of a model developed. Hence at the very beginning of the study, all elements—variables, management acts, objectives, and indicators—are jointly considered and integrated.

In addition to defining research needs and developing a model, the same or subsequent workshops can further define management objectives, construct alternative hypotheses in the model, define and evaluate alternative policies, and recognize uncertainties. In many ways, uncertainties in management objectives and in outcomes of decisions and interactions are at the source of many present environmental management problems.

But before a model can be used as a kind of laboratory world to test the consequences of these alternative policies, its degree of credibility must be explored. Note that no model—mental or mathematical—is "true." But degrees of credibility and usefulness can be defined.

There may be many ways of attempting to achieve a given objective. For example, maximum sustained yields from a fishery can be reached by controlling fishing effort through manipulation of open fishing days or by setting catch quotas.

The role of the model at this point becomes the generation of those indicators which will enable managers to decide which set of actions will best achieve the objective. Because costs and benefits arise in many forms, the manager usually needs a large number of indicators. One necessary step at this point becomes the compression of this massive amount of information to a comprehensible form. The relative merits of alternative management actions can be evaluated using the indicator output from the model. Both formal and informal evaluation techniques are useful here but, in either case, the only value is to point out policies that should be more thoroughly explored. The object is not to derive some mythical "optimal" policy, but rather to compare and combine alternative policies in order to illuminate the range and nature of available choice. Such sensitivity analyses play an important part in the evaluation process.

But methodologies are only parts of the process. Communication holds these parts together. The thick volumes which characterize the products of many impact assessment programs are an inefficient and ineffective way to communicate results. There are other ways to present the information, such as hierarchical information packages, compressed graphical configurations, and even audio-visual packages, each of which can be designed for specific audiences.

Despite the breadth and depth of these adaptive design approaches, the cost is small. An experienced core group of two analysts and two support staff could comfortably undertake one major policy design a year together with perhaps four to six preliminary "rough cut" ones. Each, of course, would draw heavily on available expertise <u>within</u> the agencies concerned with the problem. Hence the benefit is not only the policy design itself but a growing body of experience within agencies. In nearly every instance <u>existing</u> data, however scarce, are sufficient to begin, since we argue that the design of a data collection program for monitoring or baseline information must follow and be integrated with the design approaches and not precede them. Finally, the IRE's central issue for the future is how to design and plan with uncertainty. Here, at most we speculate and classify - but we do not solve.

<u>Faculty:</u>

All faculty members are jointly appointed to the IRE and another department at the university. The second department of appointment is listed in parenthesis.

F. Bunnell, associate professor (Forestry). Forestry and wildlife management, simulation modelling, field experimentation.

A. Chambers, assistant professor (Forestry). Socio-economic dimensions of resource policy, simulation modelling.

N. Gilbert, associate faculty. Statistical ecology, mathematical modelling, insect pests.

R. Hilborn, associate faculty (Canadian Department of the Environment). Systems ecology, ecological policy analysis.

C.S. Holling, professor (Zoology). Systems ecology, ecological policy analysis and population processes.

D.D. Jones, associate faculty. Systems ecology, mathematical analysis, topology.

J. Kane, professor (Zoology). Systems mathematics.

C.J. Krebs, professor (Zoology). Field experimentation, terrestrial ecology.

J.D. McPhail, professor (Zoology). Evolutionary ecology, aquatic ecology.

J. Myers, assistant professor (Plant Science). Biological control, terrestrial ecology.

W.E. Neil, assistant professor (Zoology). Aquatic ecology, laboratory and field experimentation.

T.G. Northcote, professor (Westwater Institute). Aquatic ecology, fisheries.

R.M. Peterman, associate faculty (Canadian Department of the Environment). Systems ecology, ecological policy analysis.

W.E. Rees, associate professor (School of Community and Regional Planning). Ecological and regional planning.

A.R.E. Sinclair, assistant professor (Zoology). Vertebrate ecology, field experimentation.

I.B. Vertinsky, associate professor (Commerce). Operations research policy analysis.

C.J. Walters, associate professor (Zoology). Systems ecology, ecological policy analysis.

C.F. Wehrhahn, associate professor (Zoology). Population genetics.

W.G. Wellington, professor (Plant Science). Insect ecology, bio-climatology, pest management.

N.J. Wilimovsky, professor (Zoology). Ecological policy analysis, fisheries, systematics.

ECOLOGICAL POLICY DESIGN: A CASE STUDY OF FOREST AND PEST MANAGEMENT

C. S. Holling, Dixon D. Jones, and William C. Clark

INTRODUCTION

The boreal forests of North America have, for centuries, experienced periodic outbreaks of a defoliating insect called the Spruce Budworm. In any one outbreak cycle a major proportion of the mature softwood forest in affected areas can die, with major consequences to the economy and employment of regions like New Brunswick, which are highly dependent on the forest industry. An extensive insecticide spraying program initiated in New Brunswick in 1951 has succeeded in minimizing tree mortality, but at the price of maintaining incipient outbreak conditions over an area considerably more extensive than in the past. The present management approach is, therefore, particularly sensitive to unexpected shifts in economic, social and regulatory constraints, and to unanticipated behaviour of the forest ecosystem.

Most major environmental problems in the world today are characterized by similar basic ingredients: high variability in space and time, large scale, and a troubled management history. Because of their enormous complexity there has been little concerted effort to apply systems analysis techniques to the coordinated development of effective descriptions of, and prescriptions for, such problems. The Budworm-forest system seemed to present an admirable focus for a case study with two objectives. The first, of course, was to attempt to develop sets of alternative policies appropriate for the specific problem. But the more significant purpose was to see just how far we could stretch the state of the art capabilities in ecology, modeling, optimization, policy design and evaluation to apply them to complex ecosystem management problems.

Three principal issues in any resource environmental problem challenge existing techniques. The resources that provide the food, fibre and recreational opportunities for society are integral parts of ecosystems characterized by complex interrelationships of many species among each other and with the land, water and climate in which they live. The interactions of these systems are highly non-linear and have a significant spatial component. Events in any one point in space, just as at any moment of time, can affect events at other points in space and time. The resulting high order of dimensionality becomes all the more significant as these ecological systems couple with complex social and economic ones.

The second prime challenge is that we have only partial knowledge of the variables and relationships governing the systems. A large body of theoretical and experimental analysis and data has led to an identification of the general form and kind of functional relations existing between organisms. But only occasionally is there a rich body of data specific to any one situation. To develop an analysis which implicitly or explicitly presumes sufficient knowledge is therefore to guarantee management policies that become more the source of the problem than the source of the solution. In a particularly challenging way present ecological management situations require concepts and techniques which cope creatively with the uncertainties and unknowns that in fact pervade most of our major social, economic and environmental problems.

The third and final challenge reflects the previous two: how can we design policies that achieve specific social objectives and yet are still "robust"? Policies which, once set in play, produce intelligently linked ecological, social and economic systems that can absorb the unexpected events and unknowns that will inevitably appear. These "unexpecteds" might be the one in a thousand year drought that perversely occurs this year; the appearance or disappearance of key species, the emergence of new economic and regulatory constraints of the shift of societal objectives. We must learn to design in a way which shifts our emphasis away from minimizing the probability of failure towards minimizing the cost of those failures which will inevitably occur.

The budworm case study is presented in detail elsewhere (Yorque et al., forthcoming). Rather than repeat that detail here, we shall emphasize the lessons learned as we attempted to develop and test the methodologies and concepts mentioned above. These lessons tended to expose the fairy-tale nature of many of the most treasured assumptions of ourselves and our collaborators. We have preserved the better fables to remind us of our errors, and have replaced them in practice with "counterfables." These provide a convenient focus for the discussion which follows.

DYNAMIC DESCRIPTION

<u>Fable 1:</u> Policy design should begin with an analysis of the institutional and decision environment.	<u>Counterfable 1:</u> Policy design should begin with a dynamic description of the physical and biological system.

If our goal were to analyze and prescribe for a specific problem in a specific region with specific institutional constraints, then clearly an analysis of the institutional and decision environment would have the highest priority at the very beginning of the case study. But our goal is not that. Our goal, rather, emphasizes transferability of concepts and methods to a constellation of problems occurring in various regions in various nations. It is true that to give focus we initially concentrate on a specific problem with the name budworm/forest, and on one particular regions—the Province of New Brunswick in Canada. This is just a pragmatic necessity to give concreteness and allow for testing in a real-world situation. As the steps of transfer begin to take place, it ultimately becomes necessary to examine specific institutional settings in a number of different regions and nations. This represents analysis of the implementation phase (level N - 1) suggested in Figures 1 and 2.

With our approach, in order to make transfer a reality, the initial emphasis must be on those elements of the problem which are truly general, for it will be those parts which are independent of problem, of region and of nation. This focus on generality is not possible in an analysis of institutional or decision behaviour. The state of knowledge in those fields is still primitively rooted in specific examples. In contrast, the state of knowledge of ecological systems and of ecological processes allows for well-tested analyses that have generality beyond the specific focus. With the need to facilitate transfer, the first requirement is to develop an effective and validated dynamic description of the ecological parts of the problem. At this stage, the aim is to develop a simulation model that can be used as a kind of laboratory world with some confidence that it will be responsive to the exploration of a variety of different policies and their consequences.

FIGURE 1: Elements of Ecological Policy Design

SYSTEMS LEVEL	ANALYTICAL ELEMENT	FUNCTION
N + 1	Hypothetical Overview (embedding)	Consequence check for larger societal implications
N	System Description	Specification and dynamic description of causal structure for the system under study
N	Policy Prescription	Specification of a strategic range of alternative objectives for the system and development of corresponding policies
N	Policy Evaluation	Comparison of alternatives through an array of indicators, focussing on the unknown, the uncertain, and missing components of the descriptive analysis
N - 1	Implementation	Consequence check for detailed practicality and operational feasibility.

FIGURE 2: The Process of Ecological Policy Design.

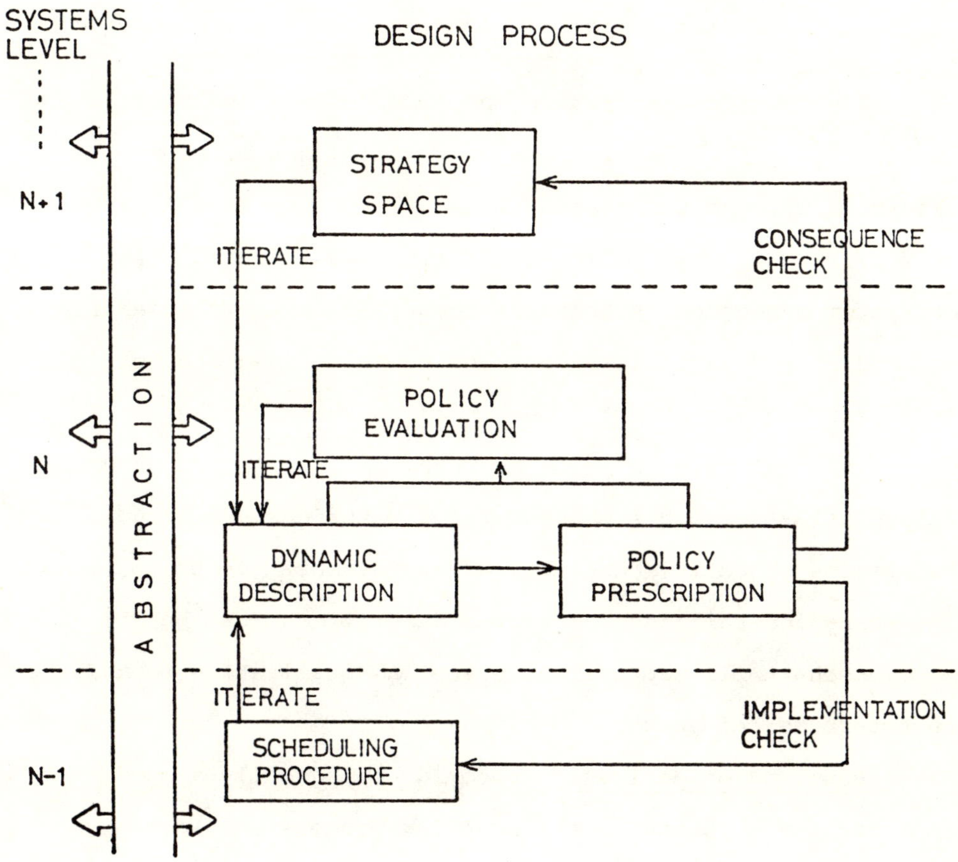

BOUNDING THE PROBLEM

Fable 2: A complex system must be described by a complex model in order to respond to complex policies.	Counterfable 2: A simple but well-understood model is the best interface between a complex system and a complex range of policies.

Any dynamic descriptive model represents an abstraction of reality. The initial steps of bounding the problem determine whether the abstraction will represent that part of reality which has relevance to the issues raised. Any ecosystem management problem is comprised of an immense array of interacting variables, conflicting objectives and competing actions. A major effort is demanded to abstract the essential elements. Our rule is to be as ruthlessly parsimonious and economical as possible while retaining responsiveness to the management objectives and actions appropriate for the problem. The variables selected for system description must be the minimum that will capture the essential qualitative behaviour in both time and space.

<u>Bounding objectives:</u> It could be logically argued that the key guidance to the bounding process should be dictated by the policy or management objectives. If the intent of the budworm case study was to analyze only the New Brunswick problem, then indeed it might be possible and valuable to start with a definition of their management objectives. But again, to re-emphasize, New Brunswick was

only chosen as a pragmatic convenience—particularly because of its troubled management history and the availability of committed collaborators in both research and management agencies. With our primary focus on transferability, an initial emphasis on the objectives of New Brunswick would have quickly constrained the range of uses and the generality of the analysis. It would have led, perhaps, to an emphasis on developing a regional econometric model and an institutional analysis with the ecological dynamics implicitly and inflexibly appended.

The budworm has evolved over the centuries to contribute to forest renewal and maintenance of species diversity. Its status as a problem depends upon the particular social and economic conditions and objectives in any region. In New Brunswick the budworm was considered only a natural curiosity until the pulp and paper industry, developed in the 1930s found they had to compete with the budworm for fibre. New Brunswick's economy is now largely dependent upon the forest industry and its objectives are social and economic.

Other regions have very different objectives. For example, the Canadian Province of Ontario has a much more diversified economy, and the forest industry makes minor use of budworm-preferred species. Their objectives relate to recreation within provincial and national parks. In the

United States, the budworm is a serious problem in Maine, where pecularities of land use and ownership make for a complex set of sometimes conflicting social, economic and environmental objectives. Europe's longer history of coupled land uses introduces still more complex objectives into its budworm problem. Moreover, even though objectives might be defined rigorously within a single region they will likely change over time. As an example, recall that the recent concerns for the environment were scarcely recognized ten years ago. Hence, we argue that an initial primary focus on objectives is inadequate to give guidance for creating a relevant abstraction.

We recognize five aspects of the budworm problem that exist to some degree in all cases. These are: social, economic, resource, recreation, and environmental. Any attempt to integrate all these into a model that would respond to a diverse set of policies would produce an analysis as complex and mysterious as the real world and useless for policy purposes. Clearly some things must be left out and our transfer goal dictates that we leave out those areas that are regionally specific. It is the resource and environmental aspects that are selected for their generality with respect to other problems, regions and nations.

These considerations of objectives, based on the need for generality and transfer, define the primary system that will be analyzed: the forest ecosystem. Any model of the forest can then be designed to generate indicators of social, economic and recreational interest which will map into the mental and mathematical models of specific regions and specific policy issues.

Bounding policies: The core of the descriptive analysis focuses on the forest ecosystem but it must, from the outset, be responsive to realistic alternative policies. The specific policies, or actions, that have been or could be applied are almost infinite—the use of insecticides, biological control agents, genetic manipulation, tree harvesting and planting schemes. Moreover, the actions which now seem to be economically impractical might, with future developments, become highly feasible. The whole range of actions feasible now and in the future fall into essentially three classes—control of the insect, harvest of the trees, and manipulation of the forest through planting. The descriptive model must allow intervention with any of these classes of action at any moment in time and any point in space.

The steps of bounding that focus on objectives and management actions must precede the actual modeling steps. Otherwise the descriptive analysis would inexorably lead to an exercise in modeling as a goal in itself. The rules for the final bounding of the problem require decisions about the number of ecosystem variables, the temporal horizon and resolution, and the spatial extent and resolution.

Bounding variables: An ecosystem of this complexity has many thousands of species and potential variables. Due to a history of extensive field and experimental studies on budworm (e.g., Morris, ed. 1963; Belyea et al., 1975), our understanding of the dominant budworm/forest dynamics is quite detailed. The system's relevant behaviour can be captured by a limited subset of variables, each of which serves a key role in determining the major dynamics of the forest ecosystem and its resulting diversity. These key variables are summarized in Figure 3.

The principal tree species are birch, spruce and balsam fir. They have a dynamic interaction of their own which is dependent on the influence of budworm. Balsam is highly susceptible to damage, spruce less so, and birch not at all. Our rule of parsimony and our strategic level of interest dictate that we only include the budworm host, balsam, as a dynamic variable.

This variable is a quantitative, or extensive measure. We must couple with it a qualitative, or intensive measure to account for tree condition. This variable is closely linked with foliage condition and retains the memory of part stress. The particular behaviour characteristics of budworm and balsam require that this variable be split into two, which we call old and new foliage in the model.

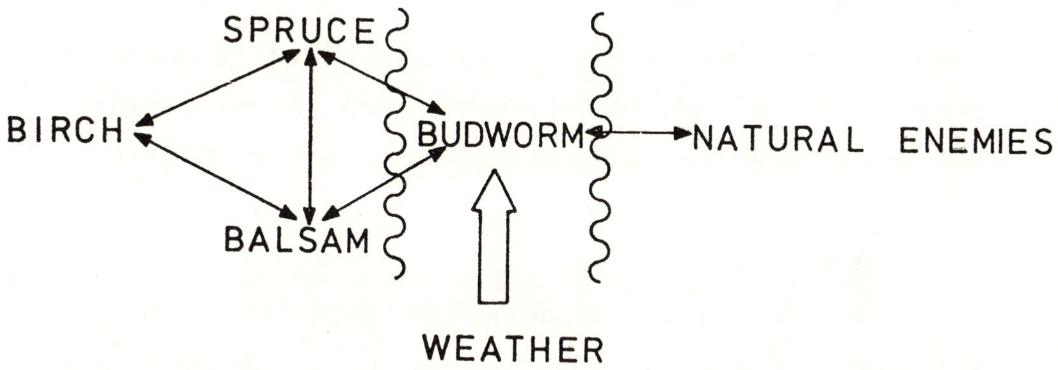

FIGURE 3: The key roles or variables and their interrelations in the natural ecosystem. The principal tree species (birch, spruce, and balsam fir) have a dynamic interaction of their own. This interaction is altered by the presence of budworm which consumes some spruce but primarily balsam. The budworm is in turn affected by a complex system of natural enemies and a stochastic weather variable. Only budworm, balsam, and weather are treated as explicit dynamic variables.

Between outbreaks the budworm is rare but not extinct, its numbers being controlled by natural enemies such as insectivorous birds and parasites. A key feature of this control is that there exists an upper threshold of budworm numbers which, once exceeded, allows the budworm to "escape" predation and multiply unchecked. Although natural enemies are an important feature whose effect must be included, it seemed unnecessary to introduce them as dynamic variables at the outset.

Outbreaks cannot occur unless the forest has recovered sufficiently from the previous outbreak to provide adequate feed and habitat for budworm. Weather is a key stochastic driving variable which can affect budworm survival, thus altering the effective forest threshold condition necessary for an outbreak.

From the thousands of potential candidates we select four as being critical dynamic variables for capturing the essential behaviour of the system: the tree host, its foliage condition, the budworm and weather.

Bounding time: An analysis of tree rings (Blais, 1968) covering eight regions of eastern North America and extending as far back as 1704 provides valuable data on the long range temporal pattern of outbreaks. These data, together with more detailed information of recent outbreaks, indicate a distinctive 30 to 44 year period between outbreaks, with occasional periods of 60 to 100 years (Figure 4). During the inter-outbreak periods the budworm is present in barely detectable densities which, when appropriate conditions occur, can increase explosively over three orders of magnitude during a three to four year period. Once initiated in a local subregion the outbreak can spread over thousands of square miles and finally collapse only after seven to sixteen years with attendant high mortality to the forest. Because of the pattern of outbreaks shown in Figure 4, the minimum time horizon required is one that can completely contain two outbreak cycles—that is, 100 to 150 years.

The time resolution that will capture the dynamics of the system is one year—the generation time of the budworm. Seasonal events within the year can be implicitly represented. This time resolution, though natural for the budworm, adds a technical complication to our representation of the forest because we must consider the age distribution of the trees. Therefore, we reluctantly are forced to subdivide the balsam variable into separate age groups.

Bounding space: The distinctive pattern in time is paralleled by one in space. Typically, the historical outbreaks spread from the points of initiation to contaminate progressively larger areas. Collapse of the outbreaks occurs in the original centres of infestation in

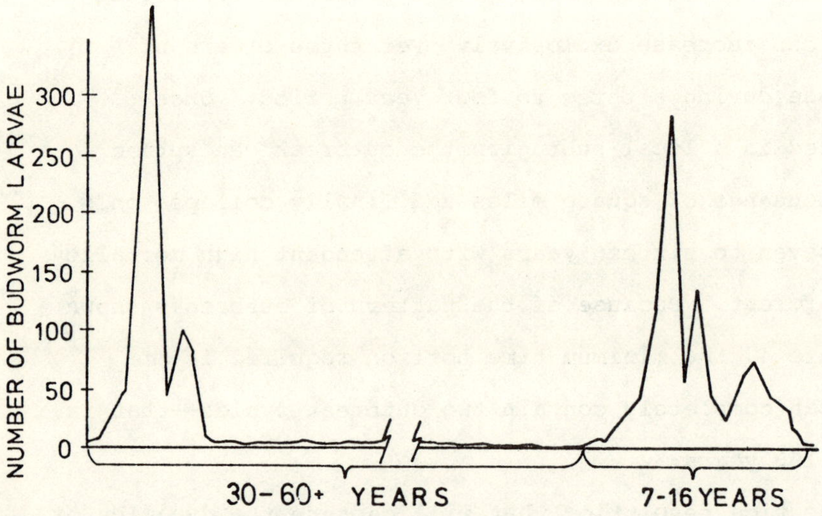

FIGURE 4: The pattern in time. Representative historical pattern of spruce budworm outbreak. There have been four major outbreaks since 1770. The density measure of budworm is what would occur on a typical balsam fir branch.

conjunction with severe tree mortality. The result is a high degree of spatial heterogeneity in forest age and species composition.

As with many pest species, the budworm has very strong dispersal abilities. The modal distance of dispersal is about 25 miles from one location, but distances of several hundred miles have been recorded. It was thought essential to have a minimum total area that would encompass at least five times this modal distance, leading to a modeled region of about 15,000 square miles. The area chosen in this study was a 17,500 square mile area containing much of the Province of New Brunswick (Figure 5). The peculiar shape is a pragmatic concession to the local management agencies but, as well, it includes the majority of the area where validation data were available. A buffer zone approximately 50 miles in width around this area compensates for edge effects.

There is high variation in the spatial distribution of the primary tree species, of harvesting activities, and of recreational potential, in part as a consequence of the historical interplay between the forest and the budworm. The 25-mile modal dispersal distance also suggests a spatial resolution less than that distance. Hence the overall area is divided into 265 distinct subregions (Figure 6), each containing approximately 66 square miles. Again the exact

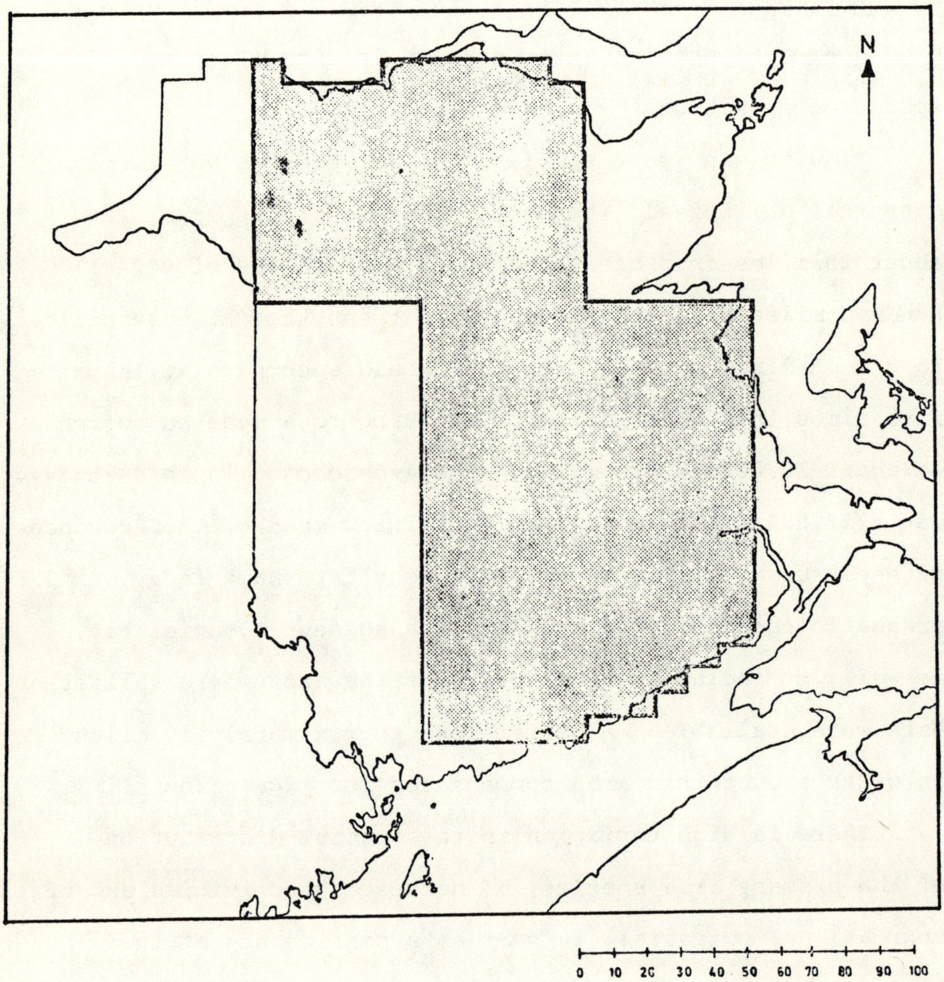

FIGURE 5: The study area used by the model in relation to the Province of New Brunswick, Canada.

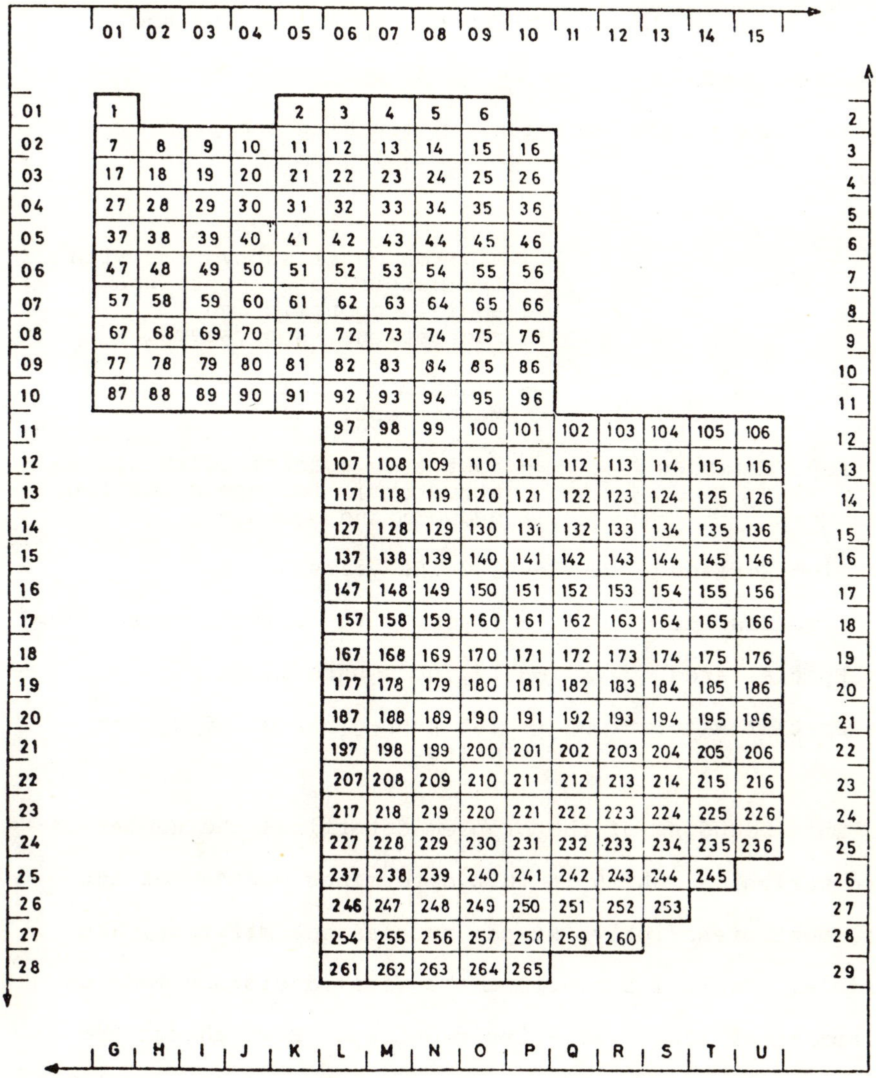

FIGURE 6 : This figure shows the numbering and indexing system for the 265 subregions, or "sites," in the study area. Each site is a bit less than 7x10 miles in dimension, including an area of about 67 square miles.

configuration is chosen to take best advantage of the validation data.

In summary, the decisions on bounding the problem are as follows:

Objectives	— models for resource and environmental subsystems with indicators relevant to the social, economic and recreational subsystems.
Policies	— budworm control and forest management.
Key Variables	— host tree species (with age structure), foliage condition, budworm and weather.
Time Horizon	— 100-150 years.
Time Resolution	— 1 year with seasonal causation.
Spatial Area	— 17,500 square miles.
Spatial Resolution	— 265 subregions of 66 square miles.

This bounding of the problem determines the number of state variables, which in turn determines whether or not subsequent prescriptive steps, such as optimization, are feasible. Table 1 summarizes the final decisions made on the number of state variables required. Even though the previous steps of bounding may seem to have led to a highly simplified representation, the number of state variables generated is still enormous. The 79 variables in each site are replicated 265 times to give a total of 79 x 265 = 20,935

TABLE 1: Number of state and driving variables per subregion.

Susceptible trees (balsam and spruce, by age)	75
New foliage	1
Old foliage (retains memory of past stress)	1
Budworm	1
Weather	1
TOTAL	79

(Other variables included implicitly)

Total number of variables in full region of 265 subregions =

79 x 265 = 20,935

state variables. Thus even this drastic simplification, accomplished through a parsimonious bounding exercise, leads to a system that is enormously complex for policy relevance. It is all the more complex since, in the above arguments, the management variables (the control variables) were not included. Highly complex models cannot be developed for highly complex purposes. Parsimony is the rule.

CAUSAL RESOLUTION

Fable 3: The goal of description is description.	Counterfable 3: The goal of description is explanation.

If description for its own sake were the only purpose, then there would be little need for a detailed understanding of causation. A multivariate statistical model would be sufficient to capture and describe historically observed patterns of behaviour. In fact that is what was done in Morris' (1963) classic study of the budworm problem in New Brunswick. The very best of sampling procedures were applied over a 15 year period in a large number of locations, and a multivariate statistical descriptive model was developed. But there are two problems. The first is that ecological systems often have key frequency behaviours that are fully represented not by years but by decades or even centuries. As already shown in Figure 4, the basic temporal pattern of this system demonstrates periodicities of 30 and more years. It is hardly conceivable that there would ever be an extensive enough range of data to allow for a full description using these methods. At best they provide an effective way to mobilize whatever data are available to point to those processes or variables which most contribute to the variance.

The second problem is that policies will develop which potentially can move the system into regimes of behaviour it has never experienced during its evolutionary history. Confidence that the predicted behaviour will be realistic in these unfamiliar behavioural modes requires considerable understanding of causation in terms of the way system subprocesses operate. A finer level of resolution in the hierarchy of causation is demanded. Yet clearly one can go too far and become encumbered by microlevels of explanation and detail that defy comprehension. Modeling at too coarse or too fine a resolution level characteristically occurs when a system is not well understood. But this need not be true of ecological systems. On the basis of a rich history of experimentation, theoretical analyses and empirical field studies the structure of key ecological processes is known not only in some detail but in a framework that has generality. This information and understanding can be aggregated to produce general and well-tested modules of key processes like growth, reproduction, competition and predation.

Consider, for example, predation. This process has been examined in great detail (Holling, 1965). It is comprised of three necessary and sufficient subprocesses—the instantaneous rates of predator attack, of competition, and of changes in predator numbers. Each of those subprocesses can be further disaggregated to their fundamental components—some of which occur universally and others which occur in particular situations only. The great diversity of predation types emerges from the many ways these non-universal components are combined.

The actions and interactions of these components have been experimentally defined and analyzed and a finite number of qualitatively distinct kinds of predation have been identified (Holling and Buckingham, 1976). For example, prey density can affect the instantaneous rate of attack in four and only four qualitatively different types. Moreover, a simple, rigorous equation has been developed whose four limiting conditions generate each of these types. Equally important, the sufficient biological conditions can be precisely defined so that the most general of information is sufficient to classify any specific situation. Such equations therefore represent the "modules" that are the building blocks for ecological models, much as an hydrologist uses the continuity equation in models of river flow.

Hence our rule of thumb is to disaggregate the model first into the constituent processes that together affect growth and survival. These processes are then disaggregated one step further into their fundamental subprocesses. The principal purpose in choosing this level of causative resolution is to increase our confidence in predictions obtained under novel policies. However, four additional and equally important benefits emerge which directly relate to our emphasis on transfer and dealing with the uncertain and unexpected.

First, transfer implies that someone is receiving the analysis. In many ecological problems the recipients include biologists and scientists with a highly sophisticated and detailed understanding of the mechanisms involved in a specific problem. Without disaggregating to the level suggested the model will, quite legitimately, not be at all credible. Moreover, there would be no way for the analysis to be responsive to the questions and knowledge that typically are focused on distinct processes.

Secondly, the organized disaggregation to the module level provides an organized way to mobilize existing data concerning partially known processes. The predation process again provides a good example. It happens that avian predators are an important determinant of the frequency behaviour of the budworm/forest ecosystem. And yet their action becomes evident only when densities of the prey are extremely low. The densities are so low, in fact, that it is impractical to sample with any reasonable degree of precision and accuracy. But once we can define the qualitative type of avian predation involved, the demands for data are dramatically relaxed. In this example the form of the equation is known with considerable certainty and only two parameters have to be estimated. Even scarce information can be assembled to, at the minimum, identify possible predator classes and then determine, through sensitivity analysis, the necessary parameter ranges within those classes.

Thirdly, modeling at this level of causation provides an effective way to deal with critical unknowns. In the example of predators mentioned above, an evaluation of alternative policies must consider their sensitivities to unexpected changes in that process.

Finally, some of the major advances in coping with the unexpected and unknown are the techniques of adaptive management (Walters & Hilborn, 1976). The key here is that, when models are uncertain, the management acts can generate information which can contribute to the understanding of the underlying mechanisms. If the models have been conceptualized at a coarse level of resolution, the experiments of adaptive management can require considerable time or extensive geographical areas to obtain results. This is impractical for management agencies with short time horizons and aversions to large scale trials. However, by disaggregating the model to the subprocess, or module, level "quick-and-dirty" experiments are immediately suggested which can yield results quickly in a localized and focused manner.

The goal then of description is not description but useful understanding.

VALIDATION

Fable 4: The purpose of validation is to establish the truth of the model.	Counterfable 4: The purpose of validation is to establish the limits of model credibility.

If the focus of interest were on developing a micro-tactical model suitable for day-by-day predictions, then a detailed quantitative validation would be demanded. But the model described here is aimed at strategic level regional planning with projections produced over large spatial areas and long periods of time. Detailed quantitative validation of such a model is not only inappropriate, it is, in one sense, quite insufficient.

The budworm problem, though prototypical in other

respects, is a rare example of a resource system with considerable amounts of quantified validation data. These data exist for each of the 265 subregions from 1953 to the present. Not all state variables were measured, but at least there are detailed insect density data. Data of this extent are rare, but even so they are still quite inadequate. They pertain only to one set of conditions: the historically managed world. During this period the system was constrained to operate within a narrow regime of behaviour, and no data are available for other behavioral modes. It would certainly be feasible, though utterly wrong, to tune the model to fit these data. Given a sufficient number of parameters, any temporal or spatial pattern of behaviour can be matched. A much more significant kind of validation has a qualitative emphasis which, despite the qualitative nature, is more demanding. The emphasis is not on specific site-by-site and year-by-year quantitative agreement for particular situations, but more on a general agreement of patterns in space and time under a wide variety of situations.

The first requirement of the qualitative validation is to match the patterns in time suggested in Figure 4. That figure summarizes extensive qualitative information concerning the behaviour of the system under no management. Under the same conditions, the model replicates with considerable accuracy this pattern, even to the point of typically generating 30 to 44 year periods between outbreaks and the occasional slip into a 60+ year period (Figure 7). Moreover, not only is the temporal pattern reproduced, but the local density changes are well within the observed range. Pattern in space is also reproduced. An example of a model run showing this spatial behaviour is presented in Figures 8a, b and c.

The second level of qualitative validation matches the patterns of behaviour with the historically managed system. In this and in all validation runs, all biological parameters have been determined by independent data and we insist they remain fixed. The only "tuning" allowed is of the initial conditions (where they are ambiguous) and the management rules (harvesting trees and spraying insecticide) applied in the simulation model. The result is shown in Figures 9a and b. The initial conditions in year 0 are set by those observed in the Province of New Brunswick in 1953. The dominant behaviour predicted is a slowly eroding forest condition and the sustained maintenance of a semi-outbreak. This is precisely what has been observed historically. The key point is that the spraying policies employed, while tending to keep the forest green and so preserving the forest industry, do so at the expense of maintaining semi-outbreak conditions, highly sensitive to policy failure.

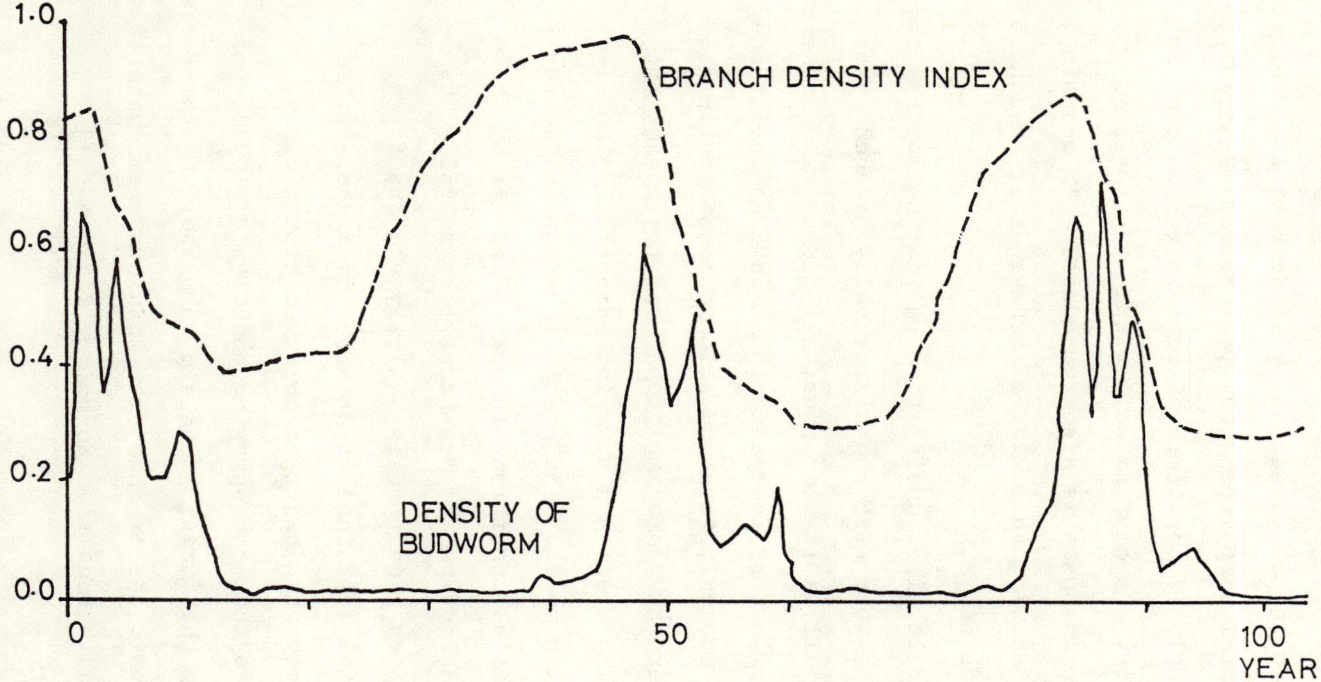

FIGURE 7: Typical outbreak pattern generated by model with no management or harvesting imposed. This represents mean conditions in 265 subregions of the simulated province, starting with initial conditions known to exist in 1953. Budworm densities are in 1000 eggs/ten square feet of branch area. The Branch Density Index is a relative scale which closely parallels average forest age and forest volume.

FIGURES 8 a, b, c:

Spatial behaviour of the budworm-forest model under conditions of no management.

The horizontal (x,y) coordinates of the figures are spatial map locations corresponding to Figure 6. The vertical (z) coordinate represents density of budworm eggs or tree volume.

Figure 8a shows, year by year, the spatial spread of a typical single outbreak. Figures 8b and 8c show the spread of, and recovery from three outbreaks over an 84-year period beginning with conditions known to exist in 1953. The typical "boom-and-bust" outbreak cycle of Figure 4 can be seen clearly.

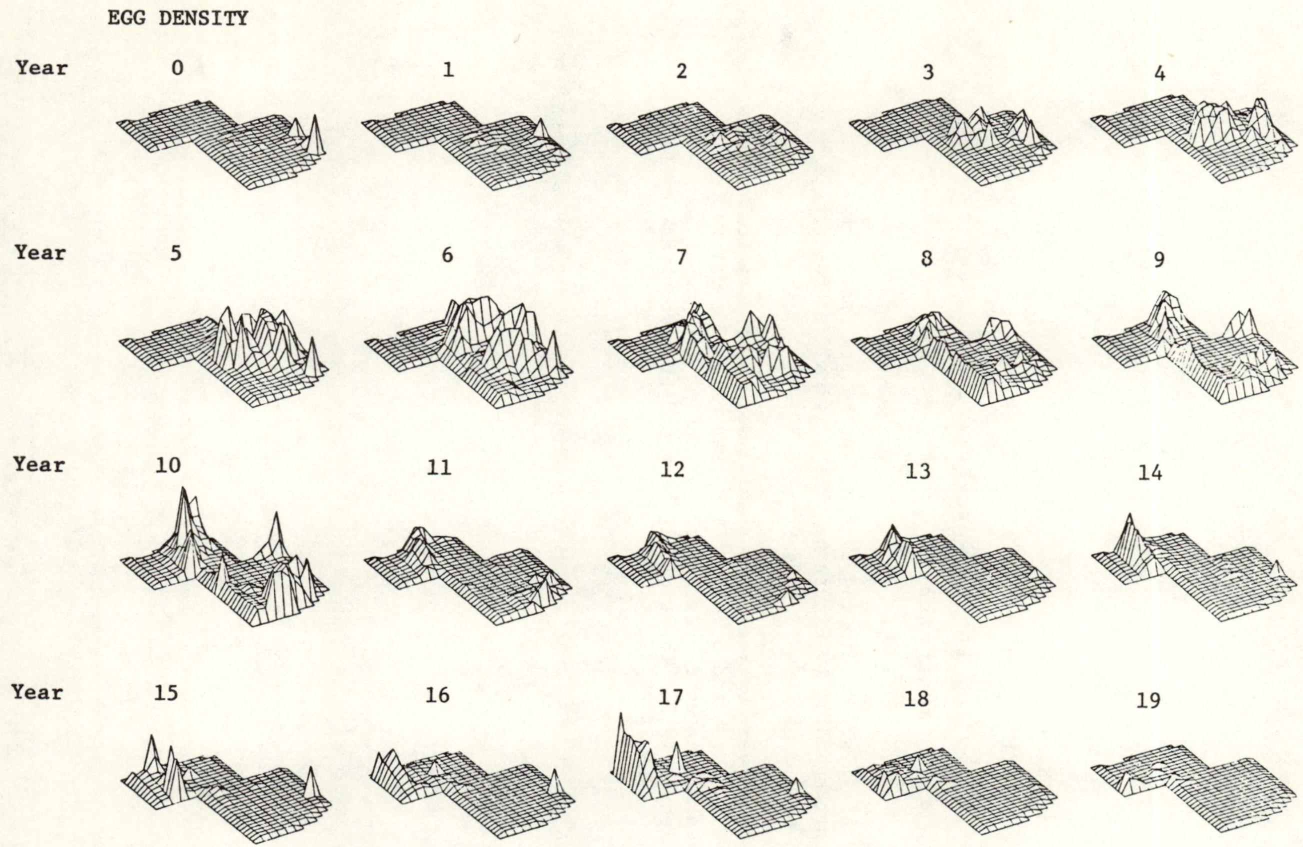

FIGURE 8a: Annual egg density patterns for a typical outbreak under no management.

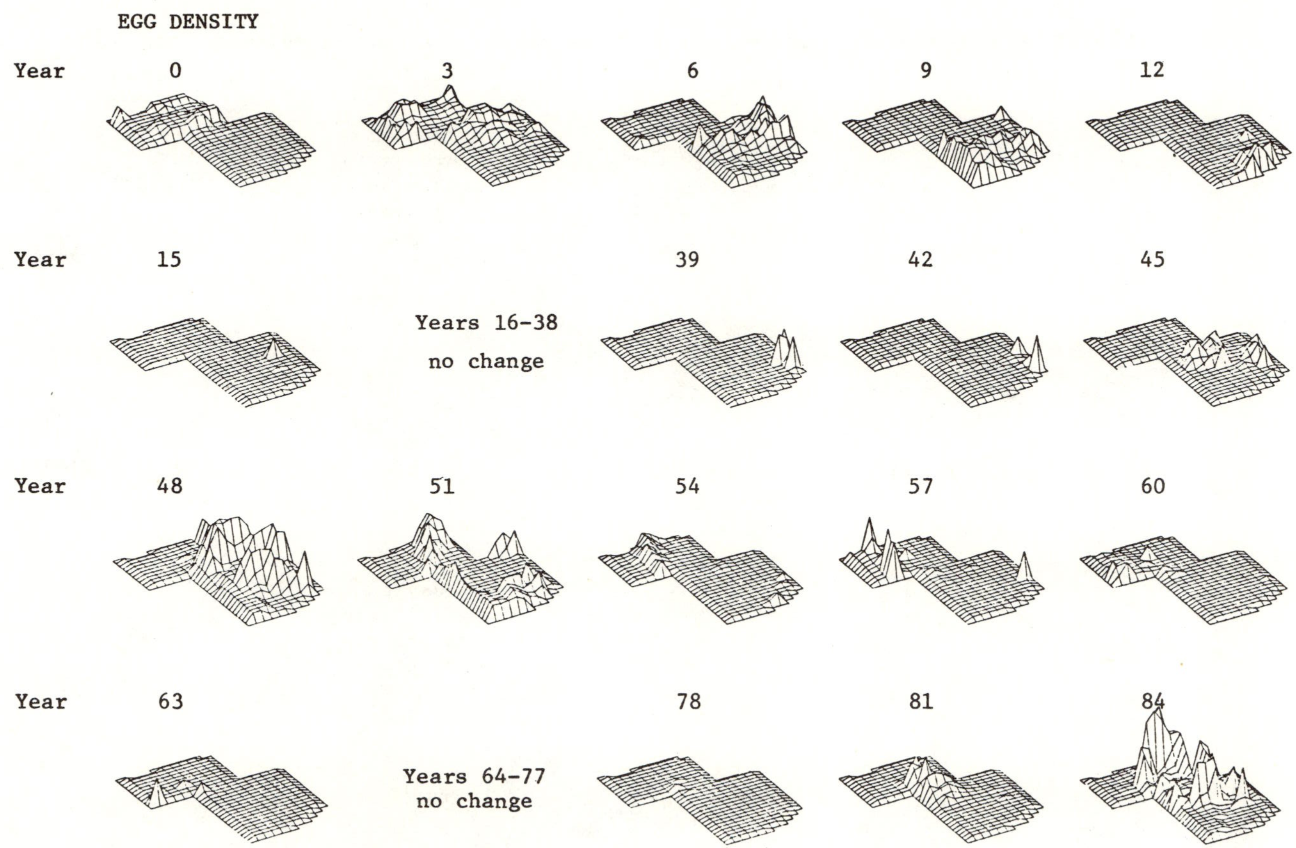

FIGURE 8b: Representative egg density patterns over an 84-year period under no management.

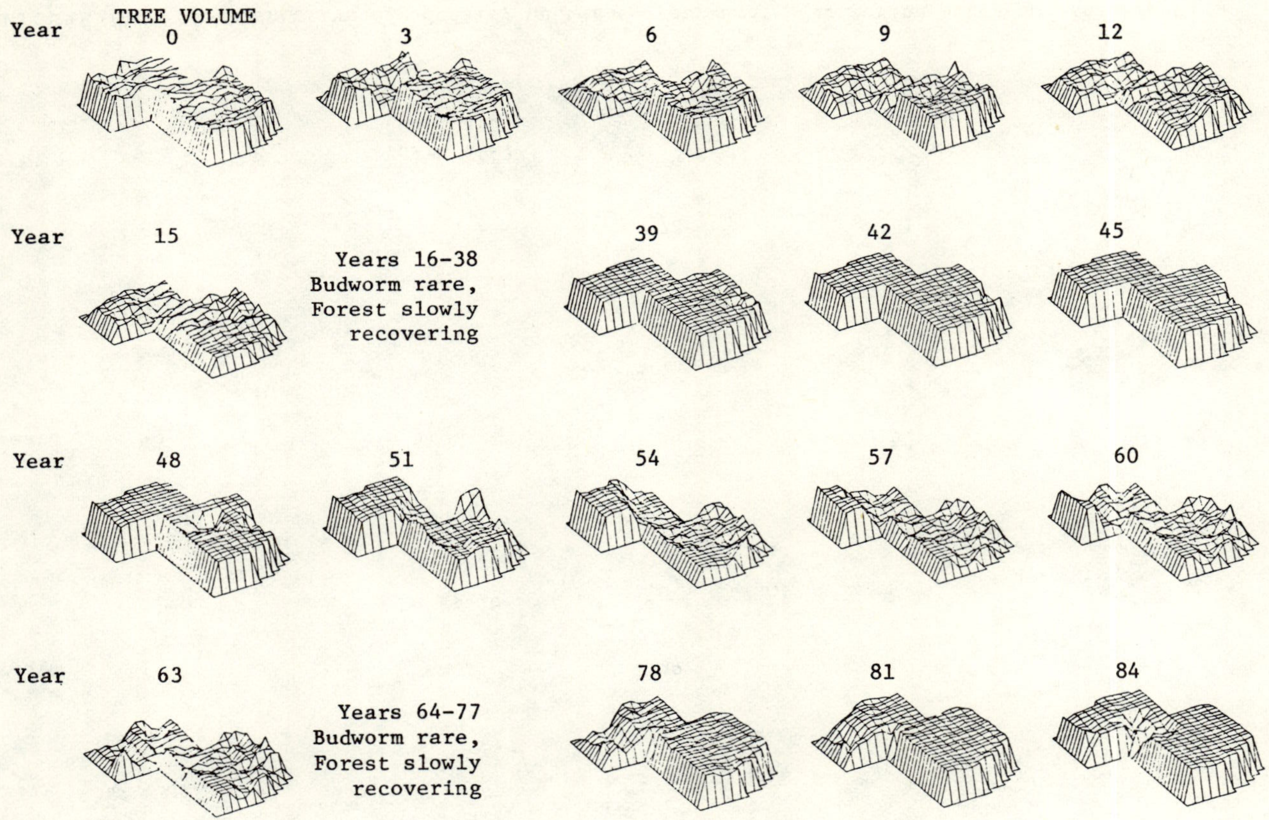

FIGURE 8c: Representative tree volume patterns over an 84-year period under no management.

FIGURES 9 a, b:

Spatial behaviour of the budworm-forest model under historical harvest and spraying rules.

The coordinates are as defined for Figure 8.

Figures 9a and 9b show patterns of egg density and tree volume, respectively, beginning with those known to exist in 1953. Compared to Figure 8, the management policies can be seen to preserve trees, but at the expense of creating permanent semi-outbreak conditions, highly sensitive to policy failure.

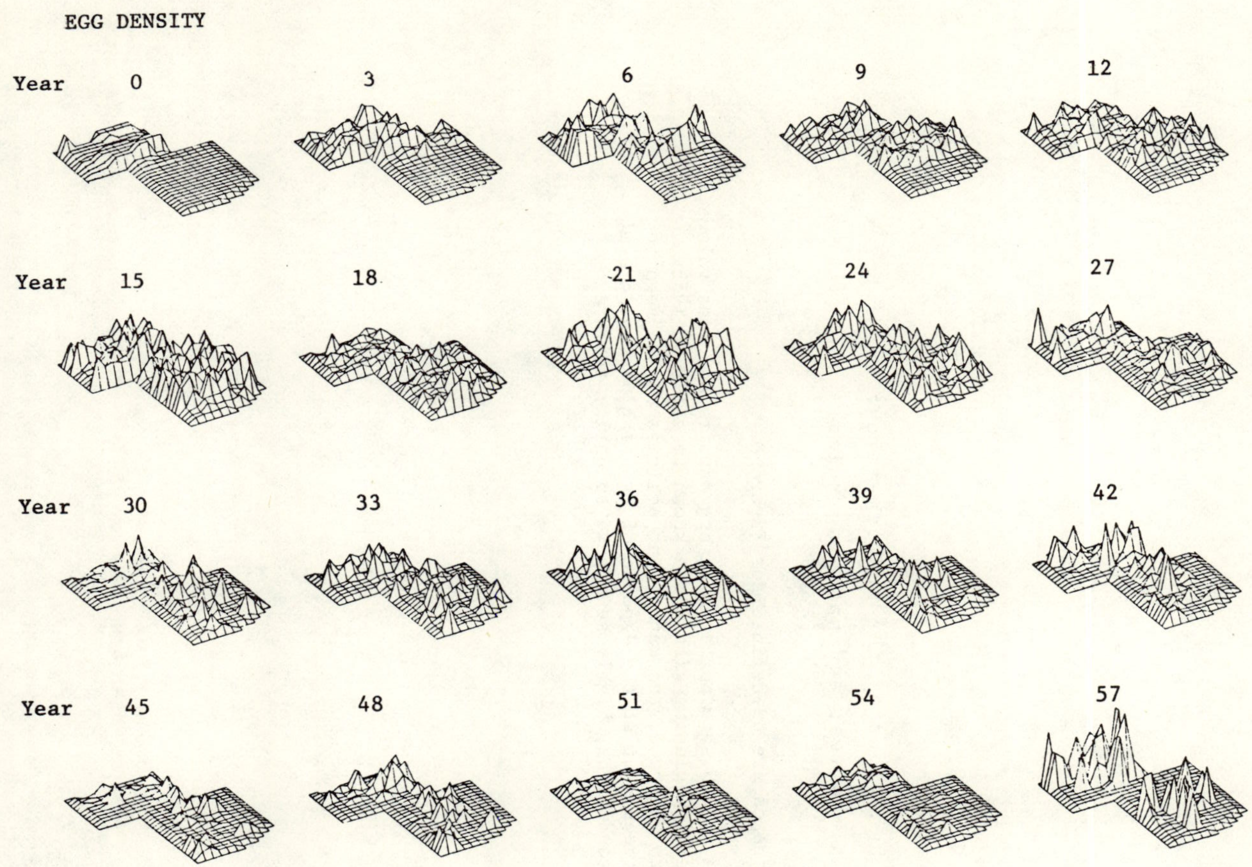

FIGURE 9a: Representative egg density patterns under historical management conditions.

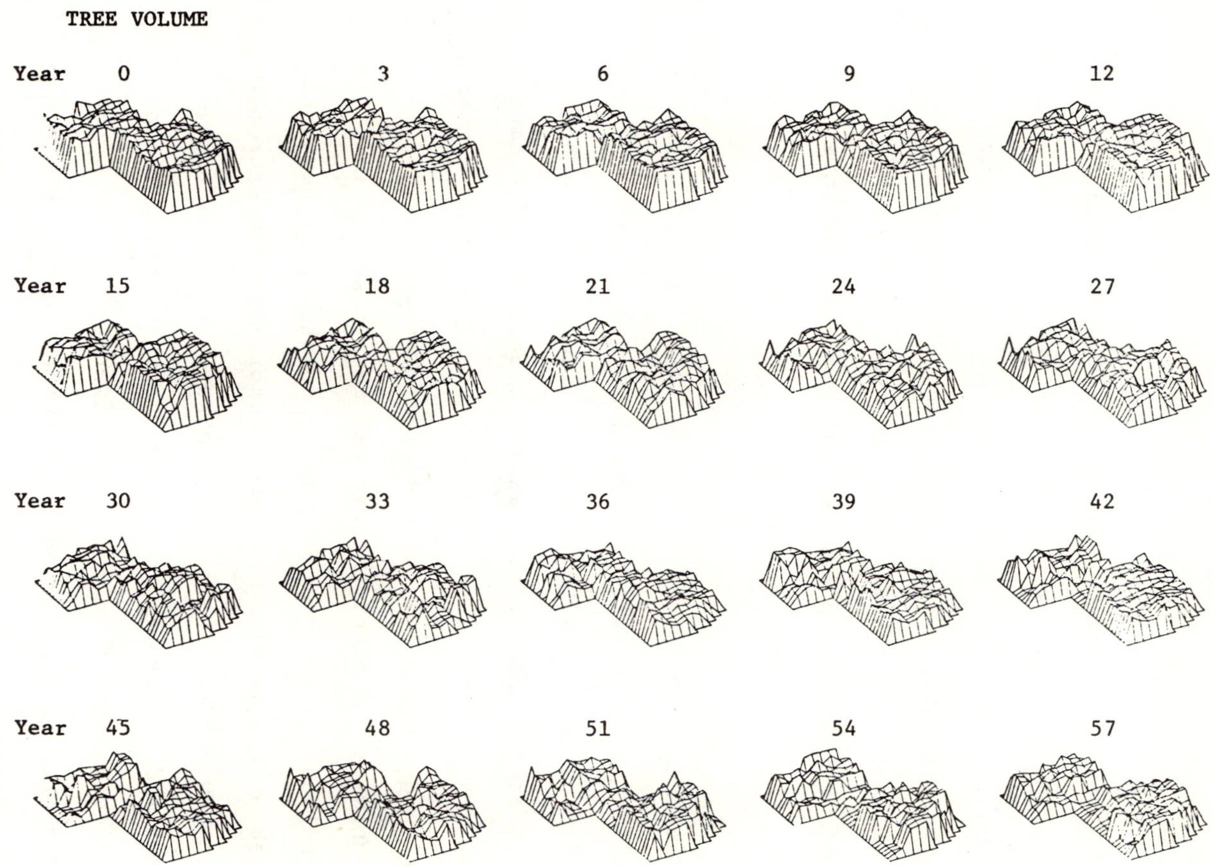

FIGURE 9b: Representative tree volume patterns under historical management conditions.

The first 19 years of this simulation run represent the period 1953 to 1975 for which detailed information is available concerning budworm densities in each of the 265 subregions. Again, the pattern agreement is striking. In both the real and simulated world the outbreak starts in the north, collapses there and throughout much of the province, re-emerges in the central regions and, towards the late 1970s, spreads dramatically throughout the whole region.

The third level of qualitative validation requires the identification of distinct patterns of behaviour occurring in different regions within the distribution of the pest. In northwestern Ontario, for example, outbreaks are more intense and tend to occur at intervals of 60 or more years, rather than the typical 30 to 44 year period observed in New Brunswick. Another pattern has been observed in Newfoundland. Prior to the recent conditions of persistent outbreak on the mainland, budworm outbreaks were extremely rare in Newfoundland. Recently, however, outbreaks have occurred and the suspicion is that they are triggered by dispersing insects from mainland regions.

The principal differences in these regions relate to weather conditions and initial conditions of the forest. In northwestern Ontario, for example, the proportion of susceptible host trees is less than New Brunswick while in

Newfoundland it is greater. Moreover, relative to New Brunswick, the weather in northwestern Ontario is more favorable to budworm and in Newfoundland less favorable. When these simple changes are introduced into the model, these regionally characteristic patterns of behaviour emerge. The model does generate periods between outbreaks under northwestern Ontario conditions of 60 years and Newfoundland has no outbreaks, unless triggered by dispersal. This kind of validation is all the more convincing because these regional differences were not known when the basic model was developed.

We are now in the process of obtaining information from Europe, where spruce budworm outbreaks are less intense and occur more frequently than in North America. Again, the expectation is that when the smaller, more scattered forest areas and the European weather conditions are included in the model, it will generate the observed phenomena. We already know from preliminary tests that the model can generate a distinct behaviour mode of moderate densities and a shorter, 8 to 12 year frequency, similar to that in Europe.

These three kinds of qualitative validation place more rigorous demands upon the descriptive and predictive capability of the model than would any effort to fit a specific time series. By focusing on patterns in space and time it is feasible to mobilize the qualitative information on a variety of behavioral modes associated with various regional conditions and historical management actions. It is this broad spectrum of qualitative matching that established our degree of confidence in a model which must explore policies that will inevitably move the system into unfamiliar regions of behaviour.

The goal of validation for a strategic model is to produce degrees of confidence that the user can subjectively weigh, as he might weigh public opinion. But a minimum is qualitative agreement of patterns of behaviour. A quantitative fit to one set of space-time data is quite insufficient.

SIMPLIFICATION AND COMPRESSION

Fable 5: The descriptive phase of applied systems analysis ends with the systems model.	Counterfable 5: The descriptive phase of applied systems analysis does not end until the systems model has been simplified for understanding.

Even the most ruthlessly parsimonious and well validated simulation model of an ecological system will be encumbered with many non-linear functional relations and many state variables. The explosive increase in the number of variables when considering spatially heterogeneous systems presents the "curse of dimensionality" in its more intractable form. Compressions and simplifications therefore are essential, in part to encapsulate understanding, in part to facilitate communication in the transfer process, and in part to exploit the potential of optimization techniques which are as yet unsuited to cope with non-linear stochastic systems of high dimensionality.

A powerful approach to this essential stage is to take a topological view of the system. This links the basic qualitative behaviour to the number and interrelation of equilibrium states. It focuses, as well, on our central concern for ecological resilience and policy robustness. Note that the model was not constructed with the initial intent of generating multiple equilibria. Rather it was based upon the detailed knowledge and data available in the literature (particularly Morris, 1963) concerning specific processes of survival, dispersal and reproduction. Nevertheless, multiple equilibria emerge as a consequence of the interaction of these processes.

This is summarized in Figure 10, where the population growth rate (the ratio of budworm population in generation t + 1 to the population in generation t) is plotted against density of budworm. These growth rate curves condense all the reproduction and survival functions within the model. As examples, when curves cross the horizontal "replacement" line (representing zero net change in population) a stable or unstable equilibrium results.

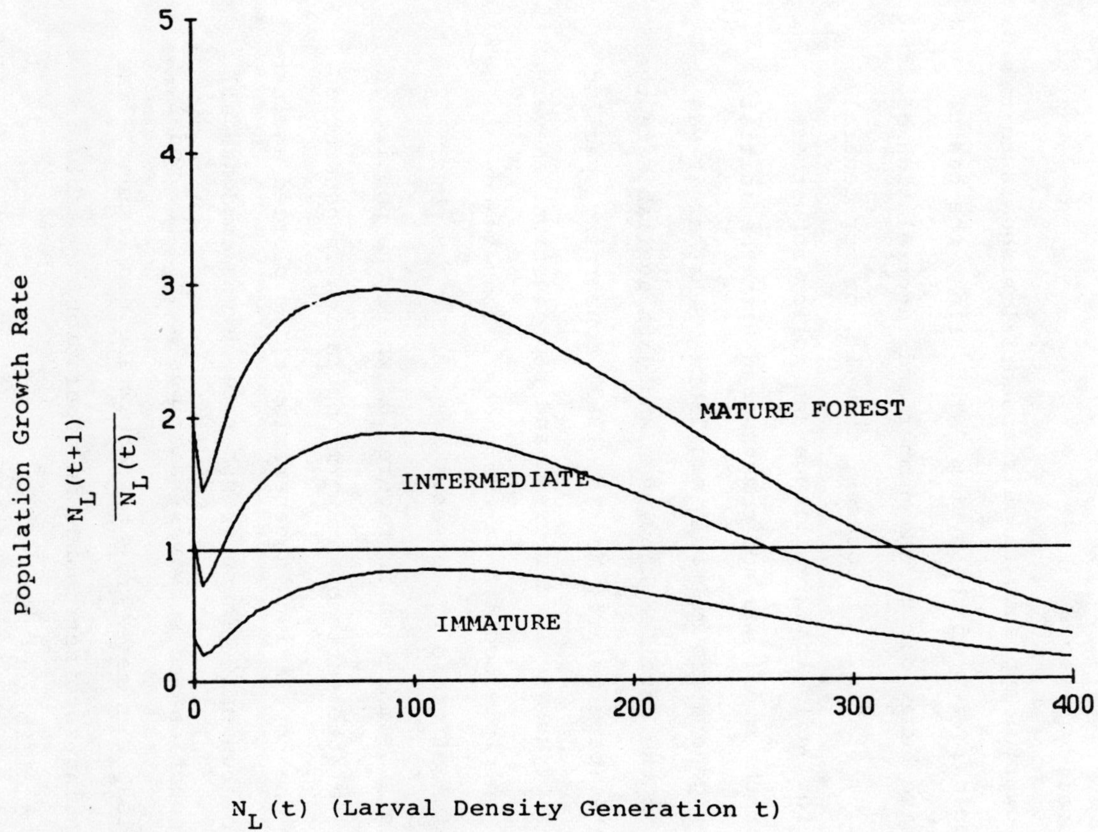

FIGURE 10: Growth rate curves for budworm populations at various budworm densities and three forest conditions. Potential equilibria occur whenever the growth rate intersects the horizontal replacement line.

The dip in the curve at low budworm densities is the result of avian predators, augmented to a degree by parasitism. When the forest is an intermediate age a lower stable equilibrium is introduced which persists until forest conditions improve and that part of the curve rises above the replacement line. An outbreak then inevitably occurs. But an outbreak can also occur by "swamping the predator-pit" through an influx of budworm from other areas. The curves generated, for this example, do not include the stochastic elements of weather which affect both survival and dispersal. When these are included, we obtain a third trigger for outbreak in the occurrence of warm, dry summers which can raise a growth rate above the replacement line.

The highest density crossover point is introduced largely through competition by budworm for foliage. Although it is presented as a stable equilibrium in this figure it is, in fact, unstable because of the response of trees. At these high budworm densities defoliation is so heavy that trees die and the forest collapses, taking the budworm with it.

A more complete and succinct summary of these multiple equilibria can be obtained by plotting all the equilibrium points in a three-dimensional space representing condensed forms of the three key variables--budworm, foliage condition and branch density (Figure 11). This represents an equilibrium

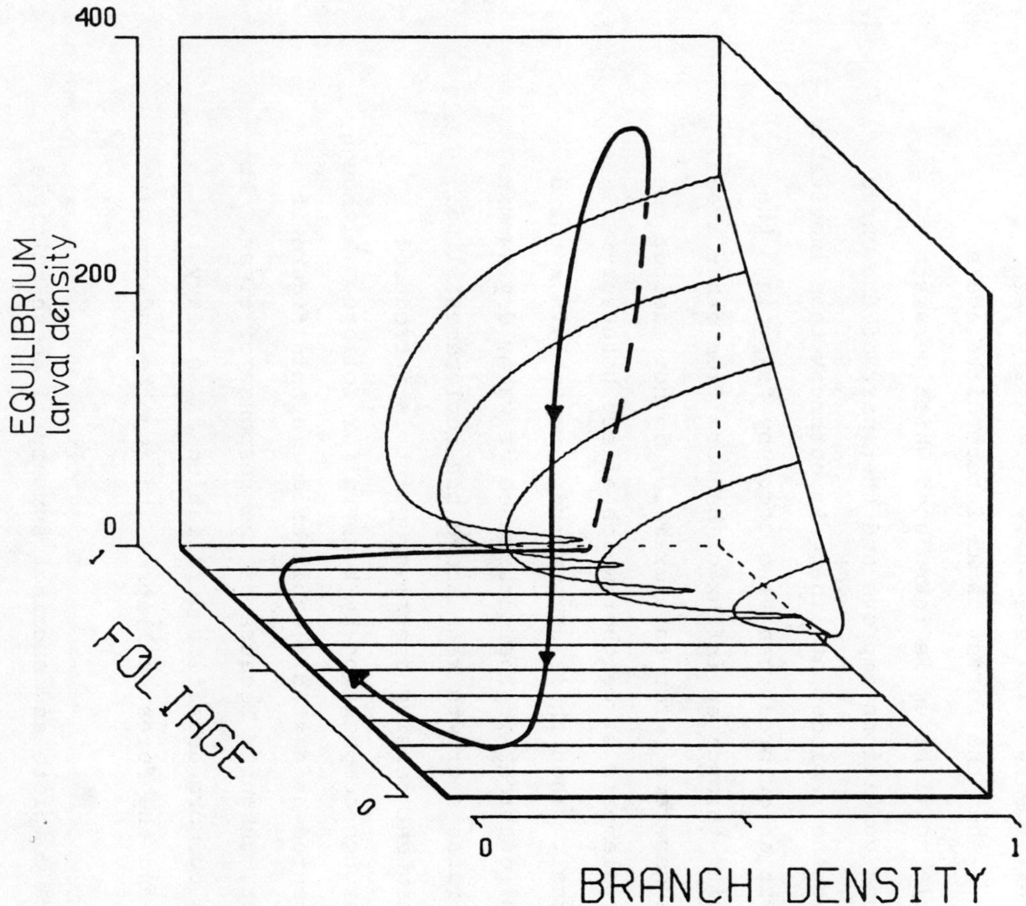

FIGURE 11: Budworm manifold (position of all equilibria levels of budworm) for different amounts of living foliage per branch and different densities of branches per acre. The trajectory shows a typical path through this space describing one outbreak cycle, in an unmanaged world.

manifold of the kind found in topology and catastrophe theory (Jones, 1976). The undercut portion of this fold is introduced by the effect of avian predators. Such representations provide a particularly revealing way of interpreting outbreak behaviour. The temporal pattern of the unmanaged system such as that shown earlier in Figure 7 can be understood by following the trajectory over this manifold as shown.

These manifold representations prove to be very helpful in condensing the simulation model. They are also a powerful device for exploring the consequence of changes in key processes or management approaches. As one example, a manifold is shown in Figure 12 in which the foliage axis is replaced by a predation intensity axis. When predation is at the level occurring in nature (1 on the scale), the "pit" responsible for the lower equilibrium is pronounced. But as predation is relaxed, the pit gradually disappears along with the reflexively folded character of the manifold. Under such conditions the whole behaviour of the system is different. A world is generated with a fairly immature forest and moderate budworm densities that oscillate on an 8 to 12 year cycle. Since insecticides potentially can affect avian predators directly through mortality or indirectly by affecting food availability, the significance of this result for management is obvious.

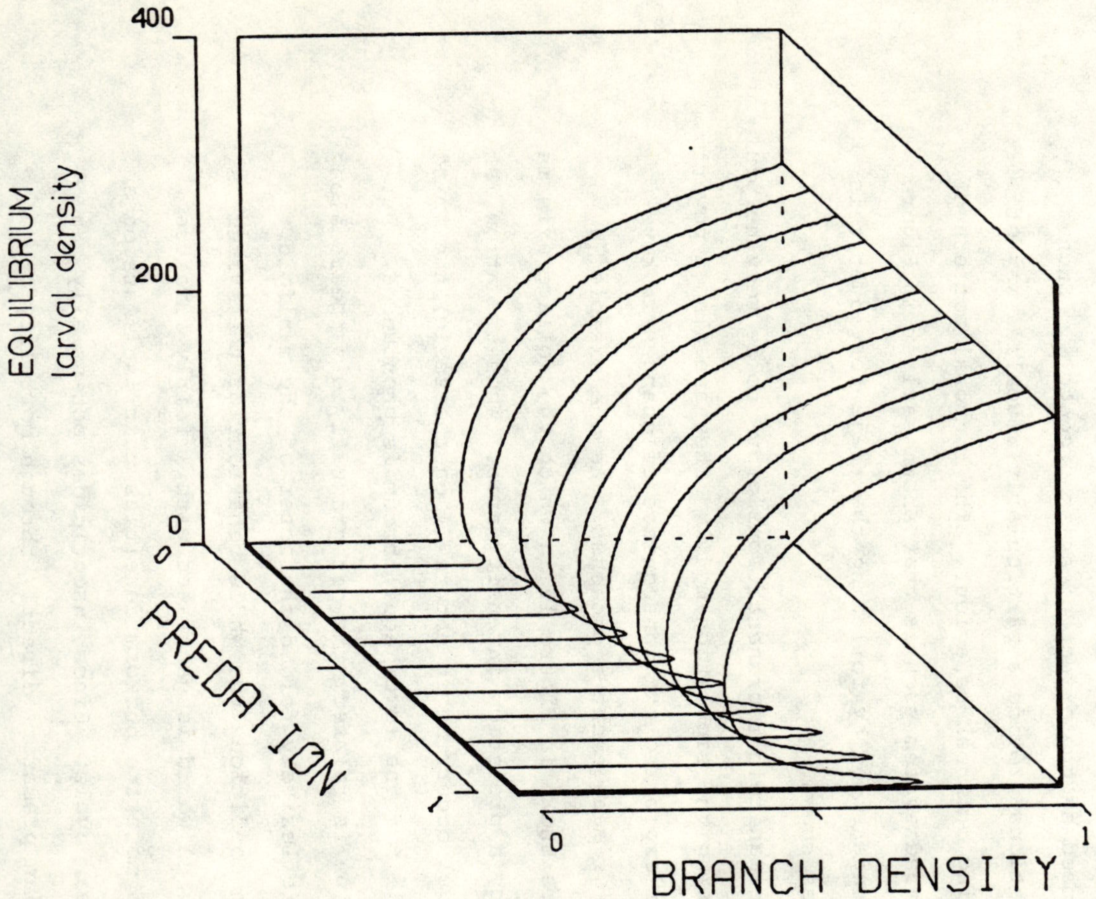

FIGURE 12: Budworm manifold at maximal foliage levels for different intensities of predation from 0 to the maximum occurring in nature (1) and different densities of branches per acre.

These manifold representations are not only useful in collapsing our understanding and providing a guide to key research and management questions, but in addition they provide a formal approach to define a small number of distinct states of the system. The budworm/forest system has eight such states which formally define various endemic, threat, outbreak and post-outbreak states. The sojourn within and between these states under various conditions can be represented as a matrix of transition frequencies, each of which has a particular benefit or cost attached to it. Moreover, as Fiering (1974) points out, such a representation also provides a succinct "back-of-the-envelope" technique for the initial development of policies.

Finally, it has been possible by concentrating on equilibrium conditions to capture the system characteristics in a small set of differential equations (Ludwig, Jones and Holling, 1977). Again the emphasis is on qualitative behaviour and powerful analytic techniques that can more definitively explore methods of spatial management designed to achieve resilient systems.

Clearly if the descriptive part of the analysis stops at the development of a simulation model, the clarity of understanding needed for transfer and policy design is seriously compromised.

ATTITUDES TOWARDS THE UNKNOWN

Fable 6: Good policy design relies upon concepts and methodologies for the rigorous treatment of the known.	Counterfable 6: Good policy design relies upon concepts and methodologies for the organized treatment of the unknown, the missing and the intentionally "left out."

Any useful analysis is based on an abstraction of reality. Such analyses therefore will always be incomplete. Attempts to "include everything" result in ambiguity, confusion, and intractability. The irony is that the more rigorous and organized the attempt to abstract a useful portion of reality for analysis, the more tempting it is to presume that those features left out of the analysis are unimportant. The more effectively the known is analyzed, the more likely it is that decisions will be based upon the analysis. But the unknown cannot be ignored and any attempt to do so is bound to end in the unpleasant surprises and policy failures discussed earlier. For effective policy design, it is therefore critically important to emphasize that what is left out at each stage of the analysis is much more important than what is kept in.

To use Walters' term, we must "look outward" from the known to the unknown. If the bounding process has been effectively accomplished then it should be clear, at least, which known systems or known phenomena have been intentionally left out. It is necessary to look outward to regions connected through dispersal or transportation processes to the managed region. Even the best of management policies designed for one region can have unexpected and disastrous consequences remote from that region. It is necessary to look outward in time as well. The budworm analysis explicitly focuses on a time horizon determined by the slowest variable in the system, i.e., tree regeneration and growth. It does not consider long-term evolutionary changes which can trigger competitive shifts in tree species composition. Similarly, short-term benefits of a management policy might be followed later by unanticipated surprises which, being unanticipated, become crises.

It is also necessary to look "upward" to those 'N + 1' level phenomena in which the detailed analysis is embedded. In the budworm study we explicitly and correctly left out an econometric model of the province and logging industry. Yet, somehow, the policies designed must be evaluated within an economic context. Finally, it is necessary to look to the variety of known, uncertain, or even hidden objectives which might be affected by decisions of management.

The methodologies associated with the looking outward approach are mentioned later when we touch on evaluation issues. Now, it is the concept which is important: an organized treatment of what is left out is the minimum requirement for a strategy of creatively managing the unknown.

OBJECTIVES IN POLICY DESIGN

Fable 7: Prescriptive analysis should concentrate upon realistic objectives.	Counterfable 7: Prescriptive analysis should concentrate upon a strategic range of individually unrealistic objectives.

The uncertainties and unknowns encountered in describing an ecological system are almost trivial compared to the ambiguities encountered in defining societal objectives. The objectives which seem so clear at any moment can shift dramatically, as testified to by the recent concern for environmental issues. Moreover, as has been discovered by the water resource planners in particular, even the best of policy analyses can founder on initially unrecognized or hidden public objectives. Since societal objectives are hidden, ambiguous, conflicting, and otherwise indefinite, the analyses rarely can accommodate them in a satisfactory manner. Hence the analyses themselves become uncomfortable, intrusive, and divisive issues of confrontation.

In response to this essential ambiguity of objectives, we felt it essential to identify a strategic range of alternative objectives containing a systematically defined spectrum of plausible and not-so-plausible management goals. Any specific example drawn from that spectrum is considered only a touchstone for the analysis and in no sense a realistic or desired objective. The goal, therefore, is not so much to define objectives which are realistic as to define a strategic range which encompasses specific objectives which may be sought by particular individuals.

At one extreme, the strategic range specifies the classical sort of unconstrained, optimally "efficient" objectives, for instance long-term maximization of expected profits in the face of known stochastic factors. At the other extreme, and equally unrealistic, are resilient and robust objectives such as those explicitly seeking the maintenance of dynamic variability. One end of the strategic range thus emphasizes a fail-safe world; the latter one which is safe in failure.

Table 2 lists eight strategically defined "touchstone" objectives explored in the budworm analysis. A corresponding range of policies was designed to achieve each of these alternatives. In an iterative process involving evaluation and comparison, these policies are now being modified, combined and refined in a realistic policy design dialogue with managers and specific interest groups.

TABLE 2: Examples of alternative objectives explored in the budworm policy analysis.

1. Retain existing management approaches ("historical management").
2. Maximize long term profits to logging industry.
3. Maximize long term profits to logging industry without exceeding present industrial capacity or operational constraints, and without violating environmental standards regarding insecticide application ("constrained profit maximization").
4. Maximize long term profits to logging industry subject to constraints of (2), simultaneously maximizing recreational potential of forest.
5. Minimize budworm densities.
6. Minimize budworm densities while eliminating insecticide applications (e.g., replacing with methods of biological control and/or forest management.
7. Transform the system's existing temporal variability into spatial variability (i.e., develop a forest in which the budworm functions as a forest manager and the essential dynamic interplay of natural forces is retained).
8. Eliminate all human intervention, both harvest and budworm control.

MATHEMATICAL PROGRAMMING AND OPTIMIZATION

Fable 8: The purpose of mathematical programming techniques is to generate optimal policies for management.	Counterfable 8: The purpose of mathematical programming techniques is to suggest interesting starting points for further development in an iterative process of evaluation and design.

Objectives—strategic or specific—specify goals. A central issue of policy design is the identification of management rules or acts (broadly, policies) which will efficiently and effectively promote those goals. We could, of course, seek to identify appropriate policies by simple heuristic gaming with a dynamic descriptive model. This is often a useful approach and almost always the best way to begin. But except in the most trivially simple cases it is a prohibitively slow, expensive, and inefficient way to develop interesting, much less optimal, policies. The number of possible policy formulations is so large that some formal guidance is necessary to define interesting regions in policy space. A variety of mathematical programming and optimization techniques have been developed to provide such guidance.

But as noted earlier, present mathematical programming techniques are just not up to the task at hand. The high dimensionality of ecological systems cripples dynamic programming, while the essential nonlinearities and stochasticities militate against such dimension-insensitive techniques as linear programming and its variants. Drastic simplification of the descriptive model is necessary to obtain any of the benefits of mathematical programming, yet with that simplification all guarantees of real world optimality for the resulting policies are inevitably lost.

Our response to this dilemma has been to employ a variety of mathematical programming techniques, not to discover the optimal policy, but rather to generate interesting probes into policy space—probes which can then be employed in conjunction with the strategic range of alternative objectives as starting points in an iterative process of policy evaluation, modification and design.

In the budworm study, Winkler and Dantzig (Winkler, 1975) used dynamic programming to calculate age, foliage, and budworm infestation conditions under which trees should be sprayed with insecticide or harvested. They resolved the dimensionality problem by viewing the forest as a collection of single trees, and handled movement of budworm between trees by assuming that the number of budworm leaving a tree would be exactly balanced by the number arriving from other trees. The analysis resulted in a set of management rules "optimal" for the extreme objective of maximizing long-term logging profits. These rules take the form of policy "look-up" tables telling the manager what to do for any possible condition of his forest (Figure 13).

It was essential to test the policies of the Winkler-Dantzig optimization in the full descriptive model in order to determine whether, in spite of the simplifications, it still provided an interesting policy for further investigation. The results were dramatic, as can be seen in a comparison of Figures 14a and 14b. The historical budworm outbreak is rapidly smothered and thereafter prohibited by the Winkler-Dantzig policy, and very little budworm-induced tree mortality occurs. But again, we emphasize that this policy must be viewed as an unrealistic but interesting starting point for further modifications, and _not_ as a "solution," optimal or otherwise, to the problem. The potential of the modified Winkler-Dantzig policy is still being explored. As one encouraging example of this potential, the system behaviour shown in Figure 14c was obtained from the policy rules, even after realistic constraints were applied to limit annual tree harvest to existing industrial capacity, to force spraying in large economical blocks rather than on a tree-by-tree basis, and to limit insecticide dosages to those permitted by legislation.

Because each formal technique of optimization forces different compromises we are also developing and applying

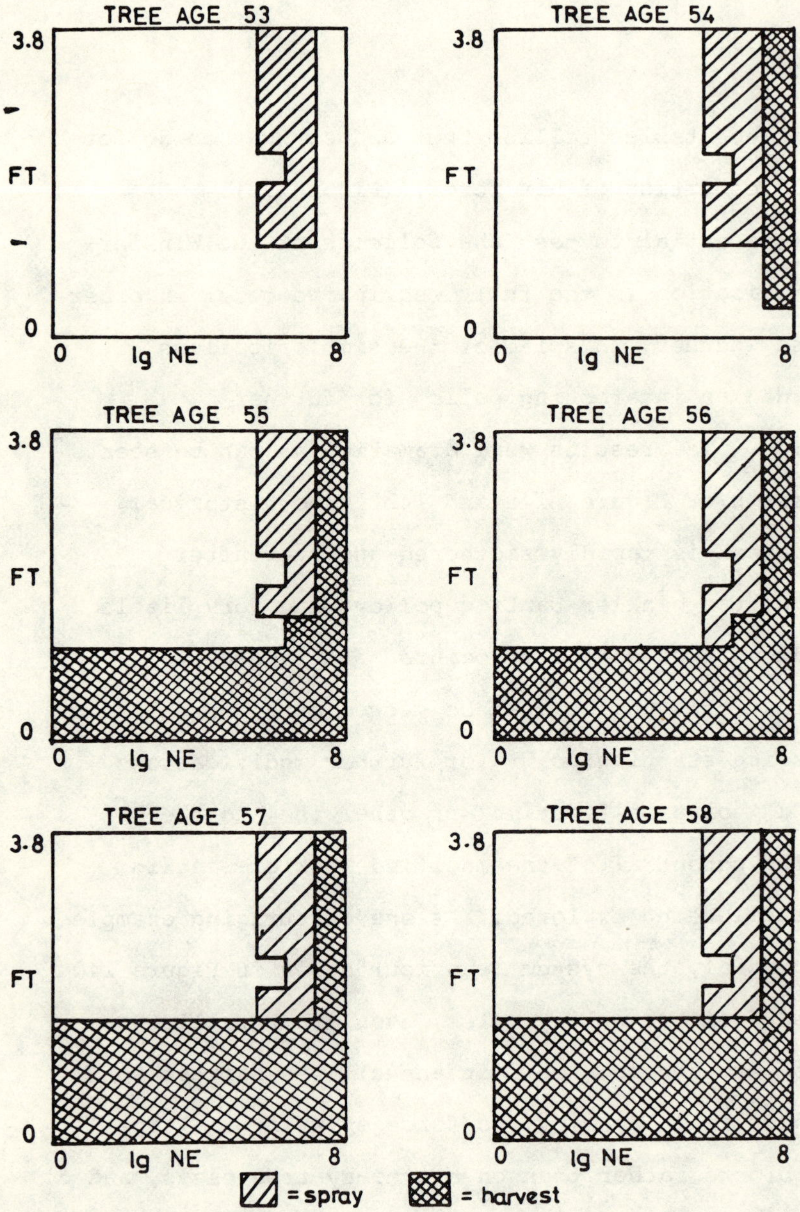

FIGURE 13: Representative policy tables generated by the Winkler-Dantzig optimization. A separate table is provided for each age of tree (or, in practice, age of stand). The table tells what management act should optimally be applied to the tree as a function of the tree's present complement of foliage (FT) and its resident budworm egg density (here plotted as lgE_D). Available management options are to do nothing, to spray, and to harvest or log the tree.

Figure 14: Behaviour of the budworm descriptive simulation under historical and Winkler-Dantzig management rules.

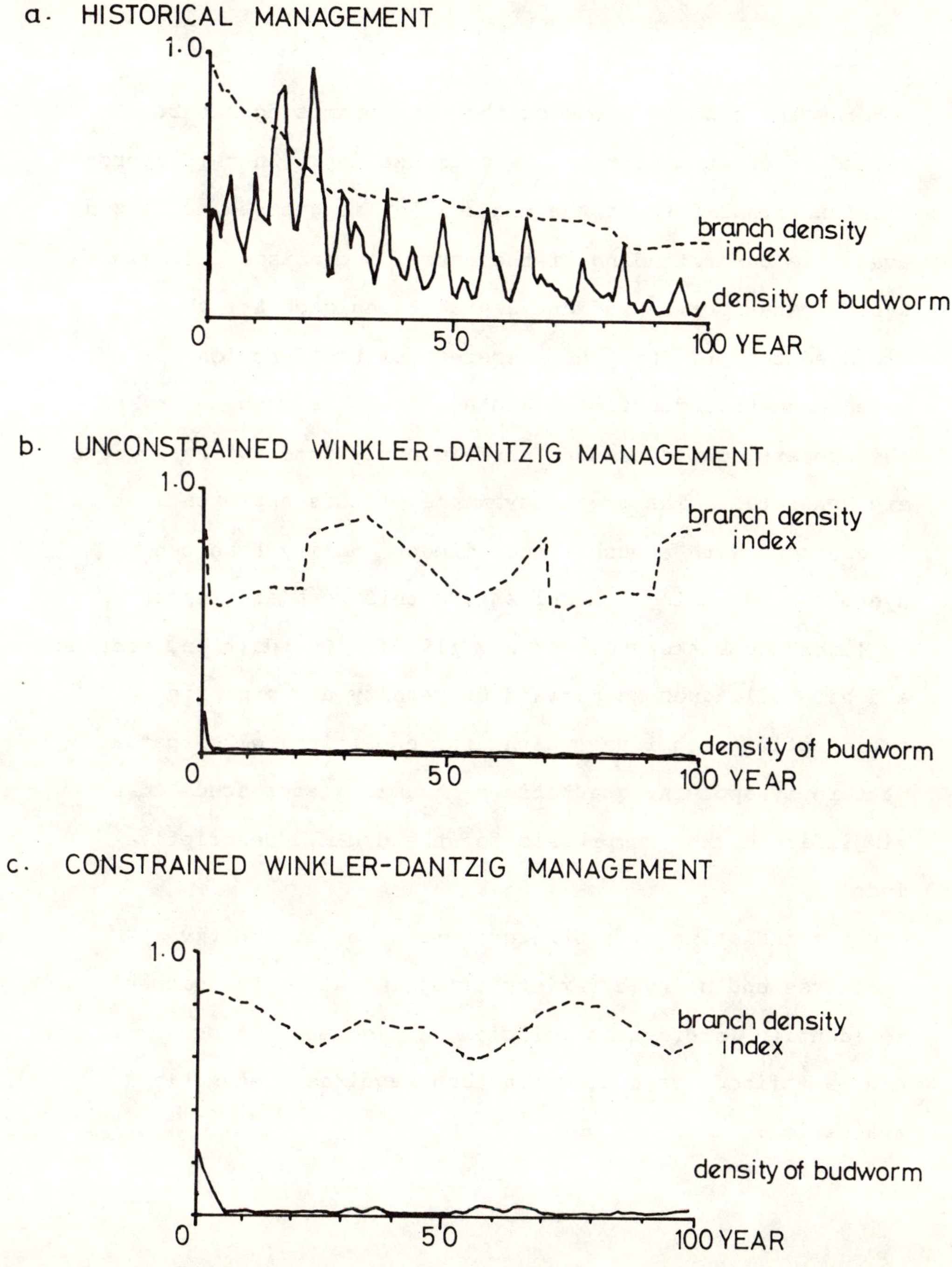

Labelling conventions are the same as for Figure 7. 'Historical' management rules are approximately those in use in 1970; 'Unconstrained Winkler-Dantzig' rules seek to maximize long term profits, but with no operational or capacity constraints; 'Constrained Winkler-Dantzig' rules introduce realistic constraints on industrial capacity and permissible levels of insecticide application.

other methodologies. One of the more promising has been termed fixed-form control law optimization. In this approach, the functional form of the control law is guessed, utilizing available understanding of the causal mechanisms determining system behaviour. Gradient search techniques are then employed to optimize the parameters of the function for a given objective function. Another guess is then taken and the process continues until sufficiently interesting policies are generated. The great advantage of this approach is that it can cope with a much higher dimensionality than can dynamic programming. Results from this work are reported at length in Baskerville et al. (1976). In addition, Fiering and his colleagues at Harvard University are exploring optimization techniques which deal explicitly with spatial pattern by applying quadratic programming approaches to a simplified Markov compression of the dynamic descriptive model.

By insisting on a strategic range of alternative ojectives and using a variety of optimization techniques to identify interesting policies, a rich menu of prescriptions can be defined, each of which then requires systematic evaluation.

THE EVALUATION PROCESS

Fable 9: The goal of evaluation is to rank alternative policies, usually by means of an objective or utility function.	Counterfable 9: The goal of evaluation is to compare and contrast alternative policies in terms meaningful to the policy designer.

Ranking implies a given set of policies, one of which must be chosen as "best" with respect to a given objective. The evaluation process properly includes such questions of choice but has a substantially broader scope. Our ultimate goal is creative design, and for this we require a rich and meaningful language to describe observed and desired policy performance. The "language" employed up to this point has been simply the state variables of the dynamic descriptive model. But socially relevant and responsible evaluations cannot be based upon state variables alone. Rather, we require a broader set of indicators relevant to those who make, and those who endure, the ultimate policy decisions. Further, it is necessary to transform the state variables into indicators in a way that explicitly reflects what has been left out and what remains unknown in the analysis, so that meaningful "handles" can be provided for the integration of other intuition, experience, and expertise available to the user.

The initial step is to develop two comprehensive classes of indicators, one focusing upon the immediate concerns of policy designers, the other on broader questions of policy resilience and robustness.

The first set of indicators is reasonably easy to generate, and can often be partitioned into categories of the sort shown in Table 3. At an early stage in the evaluation, a decision maker can choose the particular indicators which interest him and examine the time behaviour of each. There are rigorous techniques for comparing alternative policies through their patterns of indicator behaviour, and we will touch on these below. Often, however, visual inspection of the indicator graphs is sufficient to show that one policy alternative completely dominates another. This is clearly the case, for instance, when the constrained Winkler-Dantzig forest management policy (Figure 16) is compared to historical budworm management (Figure 15). Even more important, some of the original policy "touchstones" are likely to exhibit obviously desirable behaviour in a few indicators and indifferent or undesirable behaviour in others. By heuristically modifying the initial policy rules it is often possible to combine the best aspects of several policies into a composite design which satisfies most of our objectives.

The generation and examination of indicators of the known are only one part of the evaluation process, however.

TABLE 3: Examples of indicators of known interest generated by the model.

Socioeconomic Indicators

1. Profits to the logging industry.
2. Profits as a proportion of total sales.
3. Cost per unit volume of harvested wood.
4. Cost of insecticide spraying.
5. Employment rate reflecting proportion of mill capacity utilized.

Resource Indicators

1. Volume of wood in trees older than 20 years.
2. Volume of wood in trees older than 50 years.
3. Volume of wood harvested.
4. Proportion of total volume harvested.
5. Volume of wood killed by budworm.
6. Mill capacity.

Environmental Indicators

1. Damage due to visible defoliation.
2. Logging damage.
3. Age class diversity of the forest.
4. Recreational potential.
5. Insecticide impact in terms of fraction of province sprayed.

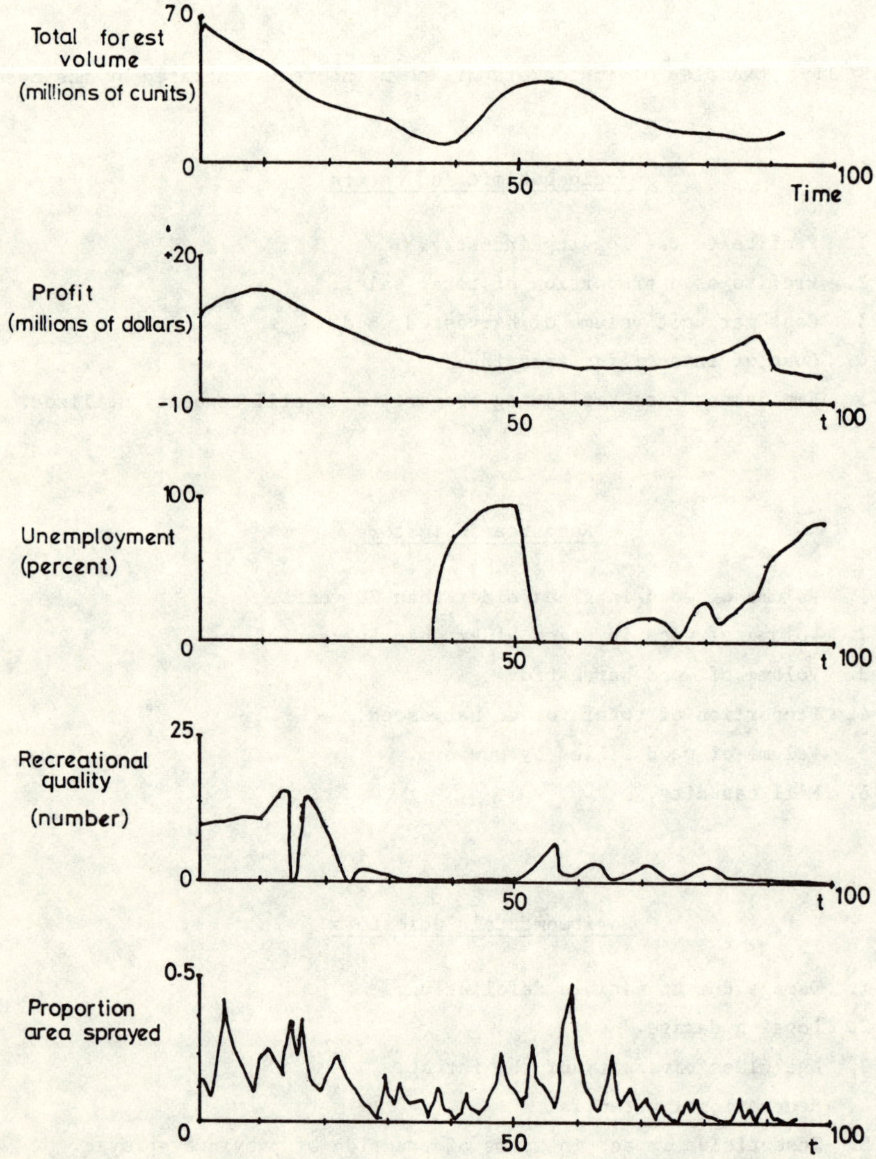

FIGURE 15: Evaluation indicators under historical budworm management.

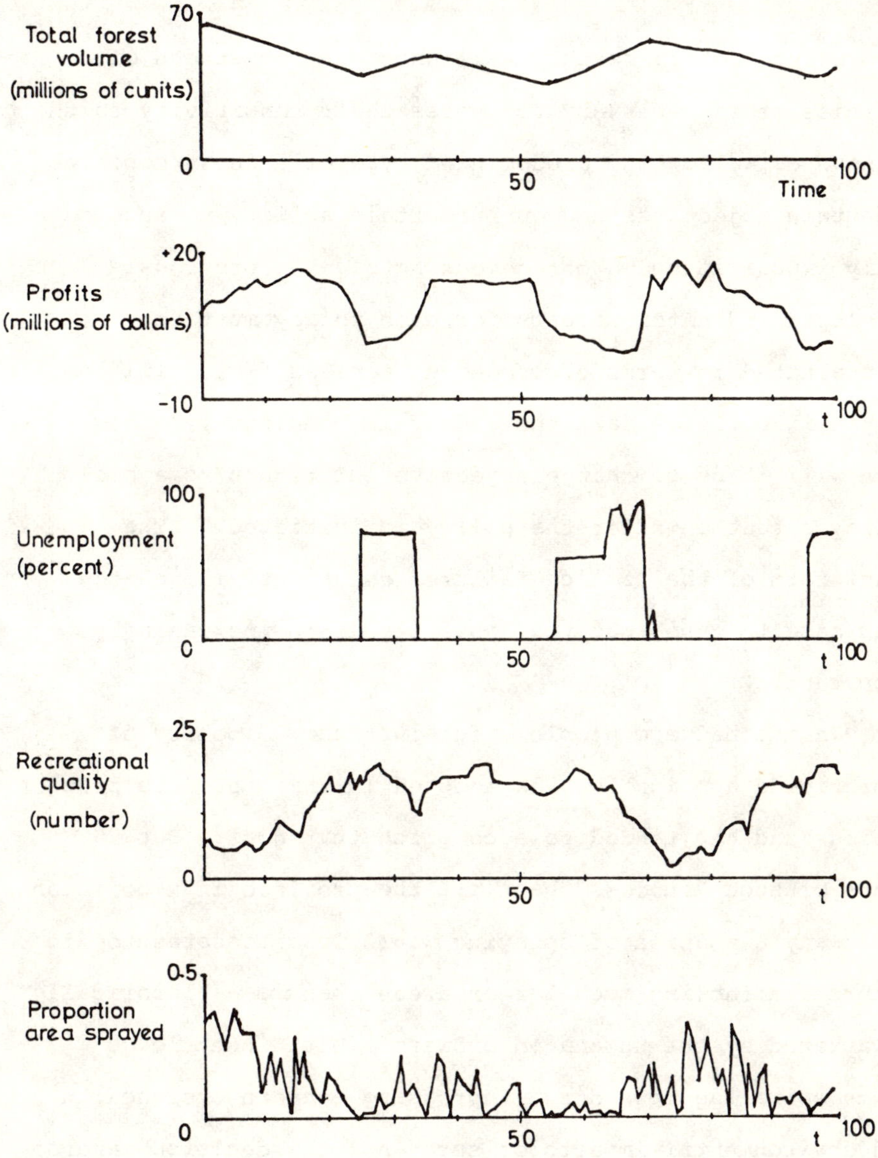

FIGURE 16: Evaluation indicators under constrained Winkler-Dantzig management.

In order to determine the resilience and robustness of policies it is necessary to assess their sensitivity to the unknown as well. One predominant type of unknown concerns uncertain objectives and our uncertain ability to successfully impose intended management acts. The previously developed indicator streams for each policy must be re-evaluated in terms of such questions as "what will happen if policies fall short or fail completely," "how hard will it be to change objectives or return to a pre-policy situation after the policy is initiated." The exact form of the "policy failure" questions will change from case to case, but the issue itself is increasingly important.

In the budworm problem, for instance, a policy of insecticide application was adopted in the 1950s to protect foliage and has tended to accomplish that goal. But 25 years of such "success" has left the province in a position where any cessation of spraying would lead to catastrophic outbreak affecting much larger areas than those historically devastated by the unmanaged budworm. With insecticide costs spiraling upward and increasing concern over health and environmental impacts of spraying, the decision makers are stranded in an impossible position with no easy options left. This sort of "option foreclosure" (Walters, 1975) surprise can and should be avoided by policy evaluation procedures.

Another important class of resilience/robustness problems concerns unknowns and uncertainties in system structure. Many of these issues can be dealt with if system description has focused on developing a topological view of system behaviour and, particularly, of equilibrium properties (cf. earlier discussion under "Simplification and Compression"). It is the number, kind and size of stability regions which determine qualitative behaviour. Shifts in qualitative behaviour have similar impact on social, economic and environmental benefits. Hence, by systematically testing the sensitivity of each policy to shifts in number and position of stability regions, measures of systems resilience emerge. And the point is not of simple theoretical interest. For example, in the Province of Quebec it recently has been observed that budworm parasite densities have increased to unexpectedly high levels. Such acute parasitism would shift the upper equilibrium of the budworm recruitment function. As a test, such a qualitative shift was introduced into the model, and led to sustained semi-outbreak behaviour over a wide range of conditions. The parasitism issue was thereby identified as qualitatively important, and steps are now underway to introduce a parasite component explicitly into the model. But the main point is that new and unexpected processes can appear, perhaps because of management. Tests of topological sensitivity provide a way to evaluate the relative resilience of alternative policies to this class of unknowns.

A comprehensive array of indicators is essential for good policy evaluation. But the more extensive the array and the greater the number of policy alternatives to be compared, the greater the danger of losing meaning in the wealth of numerical detail. For complex evaluation problems some systematic approach to indicator compression is equally essential. A number of concepts and techniques for compression in multiple attribute problems are available from the field of decision analysis, and Bell (1975b) has brought the more useful of these to bear on the budworm policy design problem.

By far the greatest conceptual and methodological difficulties are encountered in attempts to compress indicators over time. The first inclination is to employ variously weighted time averages of the indicators: means, discounted sums and so forth. But any such time averaging scheme implies a particular attitude towards intertemporal tradeoffs through which we are willing to relate the future to the present, and the ranking of policy alternatives is exceedingly sensitive to the precise nature of the attitude adopted. Clark and Bell (1976) have argued that standard market-based discounting rates are completely inapplicable to cases of ecological policy design and recommend instead an explicit evaluation of decision makers' (and, again, decision endurers') intertemporal tradeoff functions. The issue is critical and in urgent need of further study.

Even when the problem of absolute temporal compression can be resolved, however, there remains the important but generally ignored issue of local time patterns. Patterns of temporal variability are at least as significant as those of spatial variability and diversity in ecological and social systems, yet such patterns are inevitably lost in temporal indicator compressions. Bell (1975a) has developed new techniques for addressing this problem and applied them to the budworm policy design problem.

Finally, regardless of what techniques are adopted, compression is a means and not an end. Each step of compression is justified only to the extent that it truly clarifies the problems of design and choice, rather than merely simplifying them. Most compressions will properly end with the indicator array still somewhat disaggregate. The single valued utility or objective function is rarely a useful goal for the evaluation process.

COMMUNICATION, TRANSFER AND IMPLEMENTATION

Fable 10: A focus on generality and transferability lays sufficient groundwork for policy implementation.

Counterfable 10: A focus on generality and transferability is necessary for implementation, but must be complemented by a vigorous involvement of users in the design process.

We have emphasized throughout the necessity of policy design transferable to a wide variety of situations. This has been our prime motivation and justification for focusing on generality at all stages of the analysis. There are numerous advantages to this approach, but it has serious shortcomings with respect to implementation.

For implementation decisions are made in specific circumstances, not general ones. Decisions are shaped by regional constraints, by particular institutional structures, and by unique personalities. A focus on generality sets the stage for implementation but unless followed by effective application to specific situations, the analysis can become simply an academic curiosity.

Hence, close working ties have been maintained with potential policy makers throughout the design process. Three levels of transfer and implementation are presently being explored—one involving federal and provincial agencies in New Brunswick, one involving key institutions within the larger group of provinces and states affected (particularly Ontario, Quebec, New Brunswick, Newfoundland and Maine), and one involving other countries in Europe and Japan faced with similar problems. In each case the goal is not to recommend a unique policy, but rather to transfer the concepts, modeling and evaluation techniques and a menu of alternative policy touchstones into the hands of those responsible for and affected by decisions.

The emphasis throughout has been on information packages, communication techniques, and transfer workshops which can be understood, controlled and modified by the decision maker. For example, a series of integrated audiovisual packages has been prepared (Bunnell and Tait, 1974; Bunnell, 1976) to communicate as succinctly and meaningfully as possible the features of the problem, the form and philosophy of the models and the consequences of different policies (Figure 17). These are not a public relations exercise, but rather reflect our conviction that the creative communication of inherently complex ideas, stripped of their protective jargon, is as essential and challenging a part of policy design as the analysis itself. Responsible judgment by the decision maker requires understanding, not necessarily "belief," of the analysis. If this understanding cannot be conveyed, the analyst subverts the decision maker's role with no accountability for the results.

FIGURE 17: Communication and Policy Design

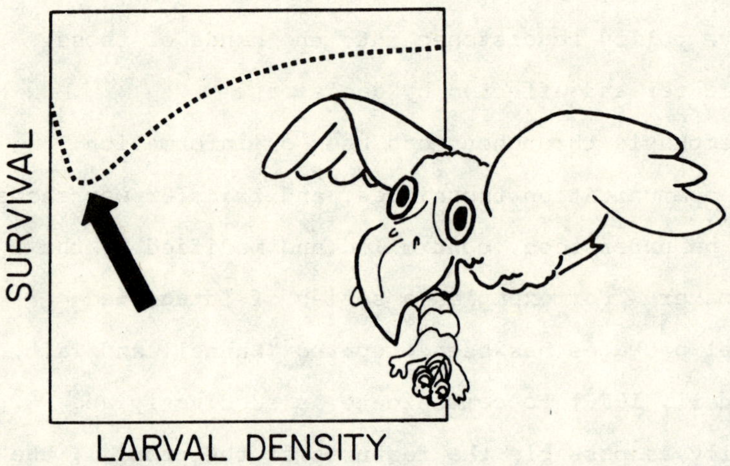

A series of integrated audiovisual packages, employing projection slides of the sort shown above, has been developed to facilitate communication and implementation of the policy analysis.

In a similar but more technical vein, graphical techniques ("nomograms") have been developed which allow visual evaluation of alternative policies via a kind of management slide rule (Peterman, 1975). Each nomogram is constructed from a large number of model simulations of different policies. The resulting display shows the effect of various intensities of cutting or spraying on a set of policy indicators selected by the user. These are presented as contour surfaces on which the manager can explore the consequences of different acts, add political and other constraints, identify tradeoffs, and begin to evolve realistic compromise policies (Figure 18). Done jointly with a number of interest groups this becomes a powerful instrument for constructive dialogue and even conflict resolution. Nomogram approaches are discussed in greater detail in the paper immediately following this.

Finally, a number of workshops focusing on transfer and implementation have already been completed: one for government and industry scientists, managers and policy people across Canada, one for scientists and provincial managers in New Brunswick, two for policy people from Ontario, Quebec, the Maritime Provinces and Maine, and one for public, industry and government groups in Maine. A similar workshops will be held for groups in Europe and Japan faced with similar agricultural or forest pest

FIGURE 18: A typical "nomogram" of the sort used in policy design dialogues with managers. (From Peterman 1977) All axes are the same, representing two management acts which can be implemented at different levels. 'Harvest Age' is the age at which a tree (or stand) will be cut; and 'Hazard Index' is an aggregate measure of insect density and defoliation stress above which insecticide spraying is initiated. Each graph represents a single evaluation indicator of the sort shown in Figures 15 and 16. Each point in the "policy space" of a given graph is the average value taken by that indicator when the system is run under the corresponding management rules. Any proposed combination of acts can be graphically evaluated by placing the cross hairs at the appropriate level and examining the contour value where they intersect in policy space. By sliding the hairs about, aggregate maxima and minima of the contour surfaces can be discovered.

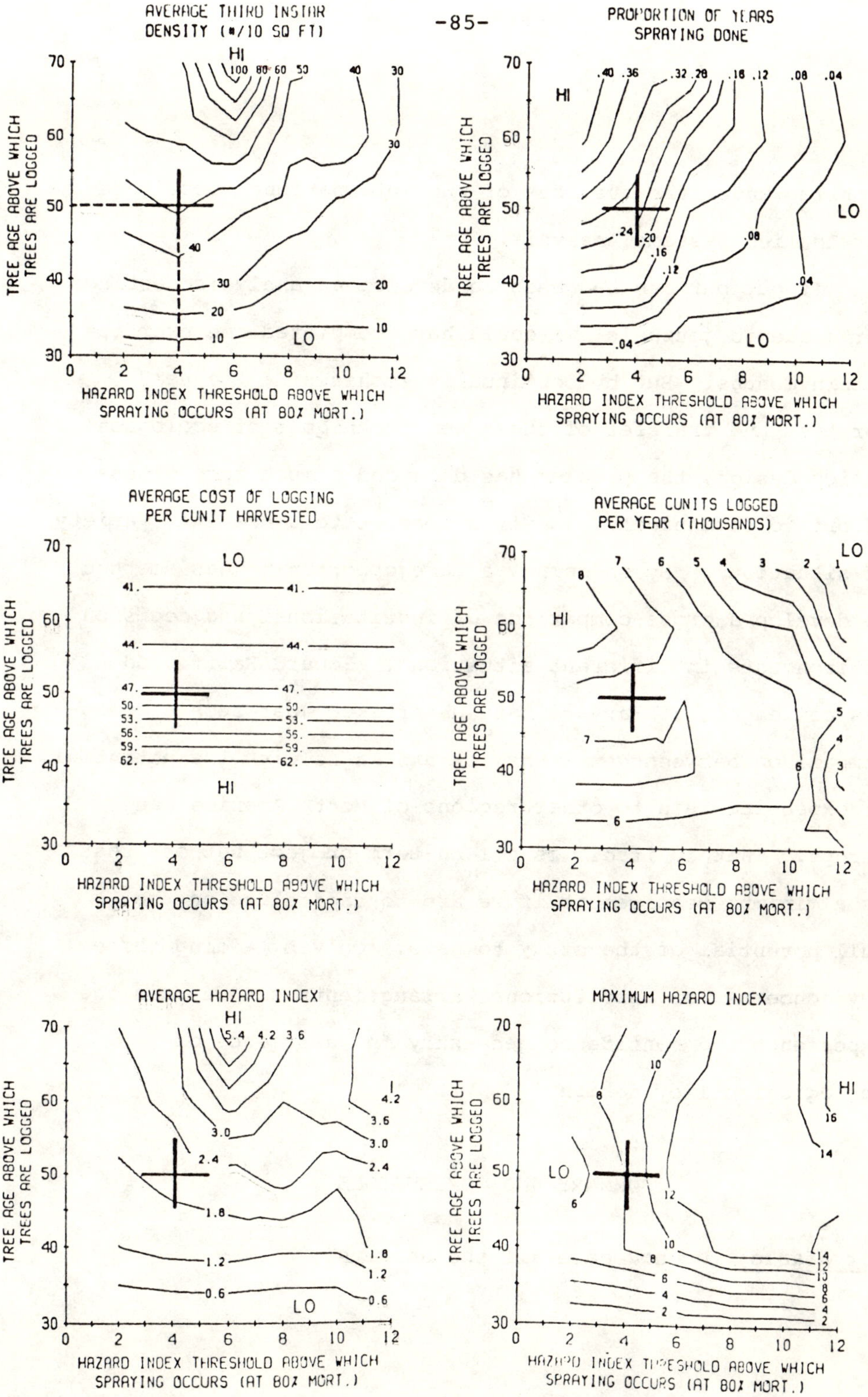
AVERAGE THIRD INSTAR DENSITY (#/10 SQ FT)
PROPORTION OF YEARS SPRAYING DONE
AVERAGE COST OF LOGGING PER CUNIT HARVESTED
AVERAGE CUNITS LOGGED PER YEAR (THOUSANDS)
AVERAGE HAZARD INDEX
MAXIMUM HAZARD INDEX
TREE AGE ABOVE WHICH TREES ARE LOGGED
HAZARD INDEX THRESHOLD ABOVE WHICH SPRAYING OCCURS (AT 80% MORT.)
HI
LO

FIGURE 18

problems under the auspices of the International Institute for Applied Systems Analysis.

If our purpose had been to develop an analysis destined for academic journals, we could have completed the exercise within months. But by continually emphasizing the need for test and transfer of these new techniques of ecological policy design, the problem has demanded a much more extensive effort to communicate, modify and adapt to users in a variety of situations. By so doing, a rare opportunity has emerged to develop a cross-comparison of institutional and decision environments in different situations. Howard Raiffa and his colleagues at Harvard have undertaken just such a cross-comparison between New Brunswick and Maine with the potential of expanding this to other regions of North America and Europe. That in itself is a long-term project but one that is a minimum requirement if we are to capitalize on the full potential of the study to date. Only by adding those key concerns for institutional arrangements can we gain the experience and confidence necessary for a science of ecological policy design.

SUMMARY AND CONCLUSIONS

Ultrafable: We now have all the answers.

ACKNOWLEDGEMENTS

In one sense it is presumptuous to thank those directly involved in this study since they were all co-equal partners in a strange inter-institutional and interdisciplinary experiment. Nevertheless, they deserve the recognition of being as much part of this creation as the authors of this paper. The policy people and scientists of Canada's Department of the Environment gave remarkable and consistent support throughout. In particular Gordon Baskerville, Charles Miller and their colleagues of the Maritimes Forest Research Centre were committed partners in the team, with their flanks admirably protected by Evan Armstrong, Dick Belyea, Murray Nielson, Dick Prentice and John Tener.

At IIASA, in its pioneering first year, an astonishing group of outstanding people gave their all to something as silly as a budworm—David Bell, George Dantzig, Myron B Fiering and Carlos Winkler.

The third institution in this effort was the Institute of Resource Ecology, University of British Columbia. Our friends and colleagues, Carl Walters, Ray Hilborn, Randall Peterman, Pille Bunnell, Nick Sonntag, and Zafar Rashid carved off pieces and solved them at times when they saw we were faltering.

Finally, and of supreme significance: Howard Raiffa, then director of IIASA, made it all happen.

REFERENCES

Baskerville, G. et al. 1976. Report of the task-force for evaluation of budworm control alternatives. Dept. Natural Resources; Fredericton, New Brunswick, Canada. 210 pp.

Bell, D.E. 1975a. A utility function for time streams having interperiod dependencies. IIASA Research Report RR-75-22.

Bell, D.E. 1975b. A decision analysis of objectives for a forest pest problem. Proc. IIASA Workshop on Decision Making with Multiple Conflicting Objectives.

Belyea, R. et al. 1975. The spruce budworm. For. Chron. 51: 135-160.

Blais, J.R. 1968. Regional variation in susceptibility of eastern North American forests to budworm attack based on history of outbreaks. For. Chron. 44: 17-23.

Bunnell, P. 1976. The spruce budworm: an ecosystem problem and a modeling approach to its management. Modular slide-tape presentation (80 min.) In preparation.

Bunnell, P. and D. Tait. 1974. A primer on models: why and how. A five-part slide-tape presentation (280 slides, 50 min.) Institute of Resource Ecology, Vancouver, Canada.

Clark, W.C. and D.E. Bell. 1976. Intertemporal indicator evaluation. IRE Workshop Paper W-7.

Fiering, M.B. 1974. Compressed policy analysis. In: Project Status Report: Ecology and Environment Project. IIASA Status Report SR-74-2-EC.

Holling, C.S. 1965. The functional response of predators to prey density and its role in mimicry and population regulation. Mem. Ent. Soc. Can. 45: 1-60.

Holling, C.S. and S. Buckingham. 1976. A behavioral model of predator-prey functional responses. Behav. Sci. 3:183-195.

Jones, D.D. 1976. The application of catastrophe theory to ecological systems. In: G.S. Innis (ed.), Simulation in Systems Ecology. Simulation Councils Proc. In press.

Ludwig, D., D.D. Jones and C.S. Holling. 1977. Qualitative analysis of insect outbreak systems: the spruce budworm and balsam. IRE Working Paper. In press.

Morris, R.F. (ed.). 1963. The dynamics of epidemic spruce budworm populations. Mem. Ent. Soc. Can. No. 31.

Peterman, R.M. 1975. New techniques for policy evaluation in ecological systems: methodology for a case study of Pacific salmon fisheries. J. Fish. Res. Bd. Can. 32: 2179-2188.

Peterman, R.M. 1977. Graphical evaluation of environmental management options: examples from a forest-insect pest system. Ecol. Modeling. In press.

Walters, C.J. 1975. Foreclosure of options in sequential resource development decisions. IIASA Research Report RR-75-12.

Walters, C.J. and R. Hilborn. 1976. Adaptive control of fishing systems. J. Fish. Res. Bd. Can. 33: 145-159.

Winkler, C. 1975. An optimization technique for the budworm forest-pest model. IIASA RM-75-11.

*In an effort to make the bibliography as current as possible, we have cited many works presently available only as publications of the International Institute for Applied Systems Analysis (IIASA) and our own Institute of Resource Ecology (IRE). These may be obtained from the following addresses:

Documents and Publications
International Institute for
Applied Systems Analysis
Schloss Laxenburg
A-2361 Laxenburg
Austria

Publications (attn. Ralf Yorque)
Institute of Resource Ecology
University of British Columbia
Vancouver, B. C.
Canada V6T 1W5

THE BUDWORM SITE MODEL

Dixon D. Jones

GENERAL

The budworm case study is part of a continuing program designed to develop, test, and implement new concepts and methodologies for resource and environmental management. The objective of this program is to link various descriptive techniques of ecological modeling with prescriptive techniques such as policy design, optimization, policy evaluation, and decision theory. Many details, particularly regarding policy, can be found in Holling *et al.* (1977) and Holling, Jones, and Clark (1976).

A prerequisite for analyzing the budworm/forest system was the development of a well validated simulation model that could be used as a laboratory world to aid in the design and evaluation of alternative policies. To be useful, that laboratory world must capture the essential qualitative behaviour of the real system in both space and time. Extensive data on budworm, forest, and economic interrelations that permit both model construction and validation have been collected by Environment Canada over the past 30 years. There are, inevitably, many missing elements, but this in fact contributes to the scope of this case study.

The full simulation model has a spatial scale of 25,850 square miles, subdivided into 393 sites of 65.8 square miles. Each of these sites uses the same program module referred to as the Site Model. The site model can also be utilized as a stand-alone entity to explore most of the full model's behaviour. It also provides a useful tool for screening trial policies.

This paper gives a brief description of the mechanisms and functions employed in the Spruce Budworm Site Model. The overall model structure is shown in Figure 1. The principal elements are a budworm survival submodel and a forest response submodel which work in parallel. Attached to each submodel is a control policy input. These controls can assume a wide range of forms and will be described in a separate document. This report considers only the part of the model contained within the dashed-line box of Figure 1.

The full model processes this module once for each site each year. Adult dispersal then intermixes the insect populations between sites. As a stand-alone entity the site model can approximate inter-site dispersal in a way that reveals some major limiting cases.

The next level of detail (Figure 2) links the major processes of the budworm and forest cycles. As this is a

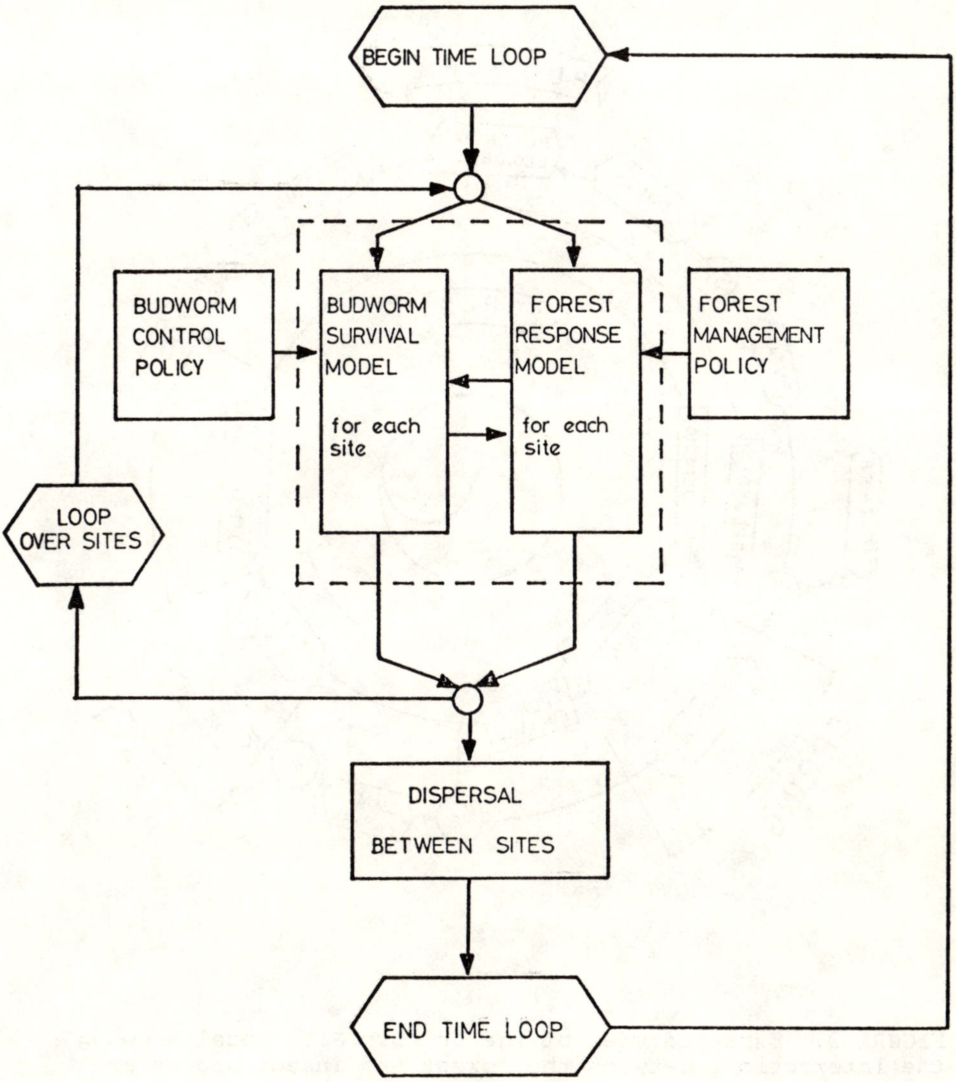

FIGURE 1: Organization of the full budworm model. This document concerns the site model -- contained within the dashed box.

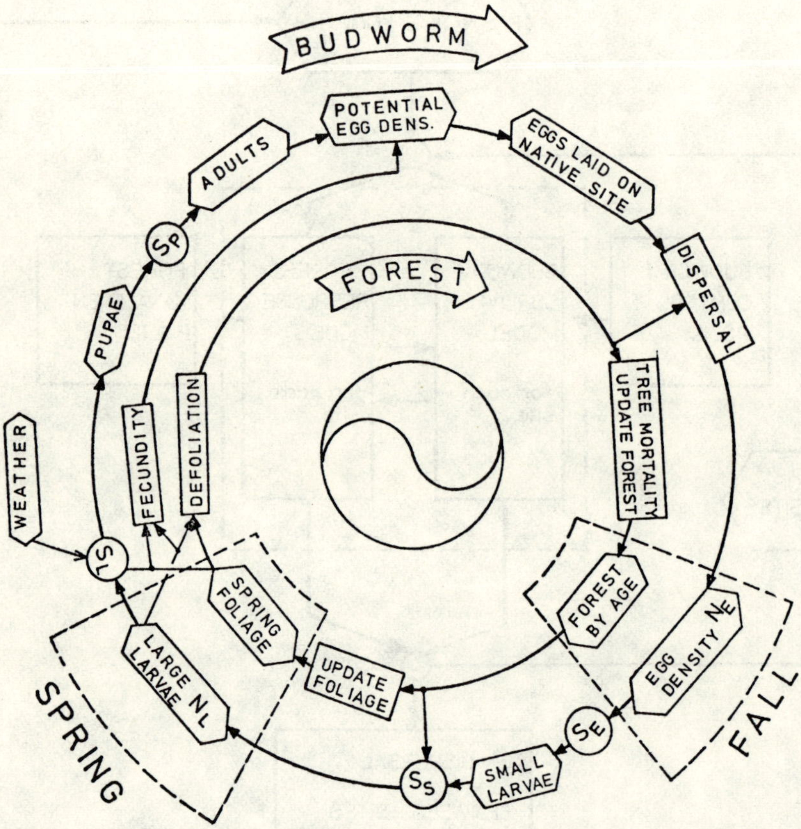

FIGURE 2: Expanded view of the Budworm Site Model showing the interactions between the forest and insect processes throughout one year.

cycle, it can begin and end at any convenient point. In the site model we choose to begin each "year" at the start of the large larval stage (Instar III). At this time the third instars have emerged and the new spring foliage has just developed. Note, however, that the full model begins and ends each year at the egg stage. This is primarily to aid in establishing initial conditions from field data for validation. It should be clear from Figure 2 that the starting point is arbitrary.

The forest and budworm submodels are described in the next two sections. The forest submodel is treated first to establish some definitions that will be needed in the budworm section.

This report is meant to describe <u>what</u> the model is rather than <u>why</u> or <u>how</u> it came to be. Those more detailed issues are placed in a separate technical appendix (to appear). Appendix A of this report gives a summary of the model equations (sufficient to reconstruct the model in the absence of the actual computer listing). Appendix B is a functionally organized glossary of terms used in the model. The symbols used in the test of this report are not the FORTRAN names of the computer programme; however, both versions are equated in the glossary.

FOREST RESPONSE

Introduction

The literature provides little guidance for constructing forest growth models for large areas. There has been considerable work on the growth characteristics of individual trees, but forest growth on a 42,100 acre site forces certain compromises with realism. The first such compromise is the assumption that the land within any site can be classified as being capable or not capable of supporting susceptible host species. In New Brunswick the susceptible host species are primarily balsam fir, white and red spruce. Because of needle morphology and timing of foliage production, black spruce is not an important host of spruce budworm. The proportion of the land on which susceptible host species grow (balsam fir, white and red spruce), though unique for each site, is considered fixed through time. This proportion is denoted as P_H (PHOST).

Within the fraction of the site that is host (P_H), we further characterize land according to age class of tree upon it. Age class as used in this model is not necessarily the chronological age but is used as a surrogate for tree developmental state. That is, trees of age class one are the understory trees at the time of overstory removal. These trees may in fact be older than 20 years but have been kept suppressed by overstory shading. Year zero then becomes the year when these trees are "released" to develop as the primary trees on their piece of land.

In this model we do not operate with the number of trees but rather with the fraction of land area occupied by trees of a specific age class. No explicit assumptions are made concerning the spatial arrangement of these age class areas. At one extreme susceptible/non-susceptible areas and age classes are completely mixed. At the other extreme each type of area is a contiguous subregion of the site. The model does not distinguish these extremes. The model parameters define an implicit spatial distribution that is characteristic for sites in New Brunswick. An explicit statement of what this distribution is and the effects of changing it will require a supplementary stand model using a much finer spatial resolution.

Tree Age Structure and Growth

The forest model used does not deal explicitly with trees, but rather with the total proportion of land associated with each age class. A schematic description is illustrated in Figure 3. Suppose that on a given plot of land there are some host trees one year old, some two years old, some three years old and some land with no host trees at all. The natural mix of these types of land might

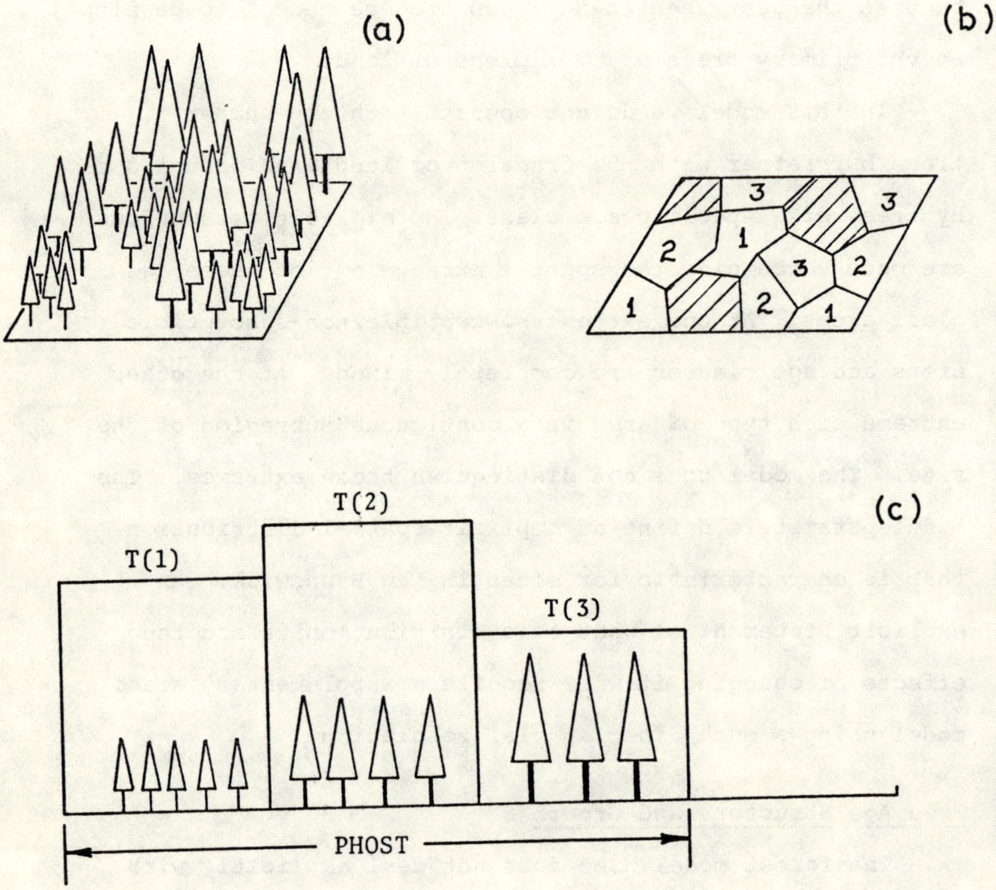

$$\sum_{i=1}^{3} T(i) = 1$$

FIGURE 3: Schematic view of the forest variable (see text).

be something as shown in (a). Rather than count the number of trees in each age class, we subdivide the land into areas associated with each age class. We have done this in (b) by dividing the plot into several polygons, each containing only one age class. In the limit we may have to draw polygons around individual trees, or, as is more likely, around small groups of trees of the same age. The basic simplification concerns the overlap of inter-crown mixing between age classes. We assume this effect is small and is uniform for any pair of neighboring age classes. We now add all polygons of age class one and call its fraction of all host land T_1. We do the same for age classes two and three to give T_2 and T_3. The aggregate is a tree age vector, or histogram as shown in (c). This rearrangement aids in visualization of the bookkeeping procedures, but it does not imply that the actual forest is operating as single age stands.

The budworm model uses not 3 but 75 age classes. The vector element T_i is the proportion of host land in age class i. The last age class, T_{75}, is the proportion of host land belonging to age class 75 plus all older age groups. The tree age vector is normalized so that each element, T_i, is the proportion of host-covered land rather than the proportion of all land. That is, $\Sigma T_i = 1$ and not P_H.

Once each simulated year, the value in compartment T_i is advanced into the compartment T_{i+1}. Compartment T_{75} is not advanced but accumulates quantities from T_{74}. In addition to ageing, land can be removed from compartment T_i through mortality to the ith age class. There are three forms of mortality: logging, budworm induced and "natural." Natural mortality affects only T_{75} at a rate of 2.37% per annum.

When trees are removed for any reason, that portion of land is reallocated to compartment T_1. Thus, total mortality equals yearly regeneration. A diagram of the bookkeeping procedure for tree age classes is shown in Figure 4. We assume that all land has an understory (equivalent to T_1 type trees) that is released whenever the older trees are removed. We also assume that the understory is not affected differently by different mortality agents (logging, budworm, or old age). However, it would not be difficult to change this assumption and, for instance, revert logged land back to a younger age class than budworm damaged land. We have not yet defined precisely what a "year-one" tree is. We do this in the next section.

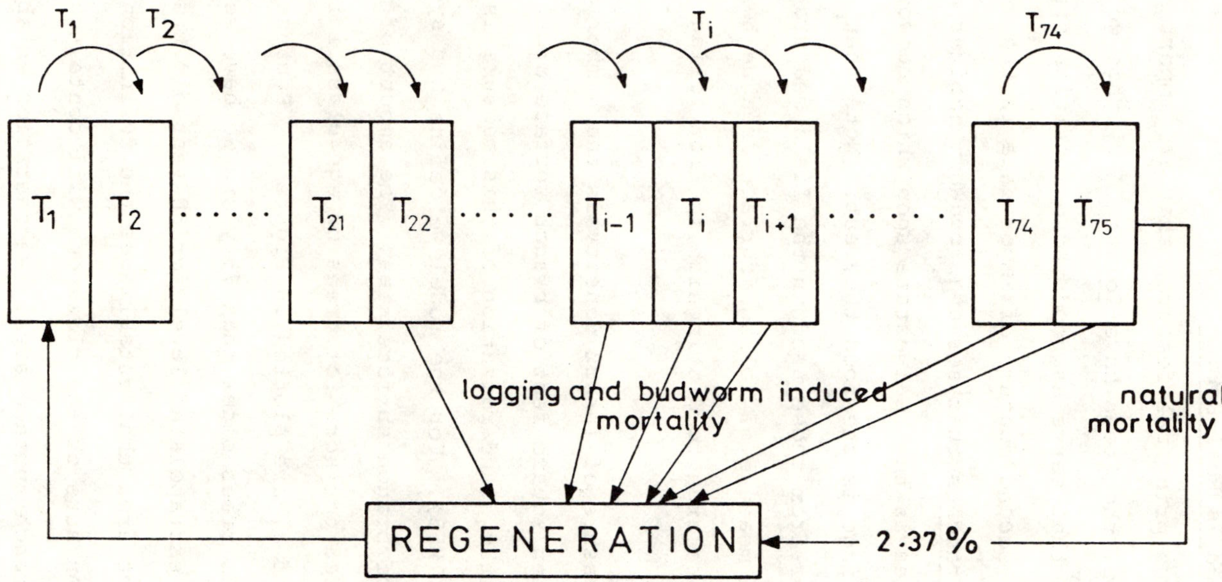

FIGURE 4: Schematic view of the tree age vector showing the method of updating and removal by budworm, logging, and "natural" causes.

Branch Surface Area as Real Estate

Trees have two principal features that are important to budworm. First, the trees have foliage. By this we mean the green needles that serve as food and oviposition sites for budworm. The second feature of importance is the surface area of branches that serve as budworm habitat. Branch surface area is the real estate upon which budworm reside, while foliage is the food they eat. Note that these terms differ from those used in the field measurements. The field measure of "foliage surface area" we here call branch surface area to distinguish it from foliage as a food resource.

The field measure of insect density is the number of individuals per ten square feet of branch surface area. A description of the standardization of this measure is given in Morris (1955). For the model we define the symbol tsf as the unit of branch surface area. The amount of branch surface area per acre for trees of age class i is σ_i. The values of σ_i are plotted in Figure 5. Because age group T_{75} contains trees older than 75 years we have set σ_{75} equal to the extrapolated level for 80 year old trees.

It is surface area that establishes the definition for age class one. An acre of age class one will contain 8750 square feet of branch surface area. Regeneration establishes trees with this branch density.

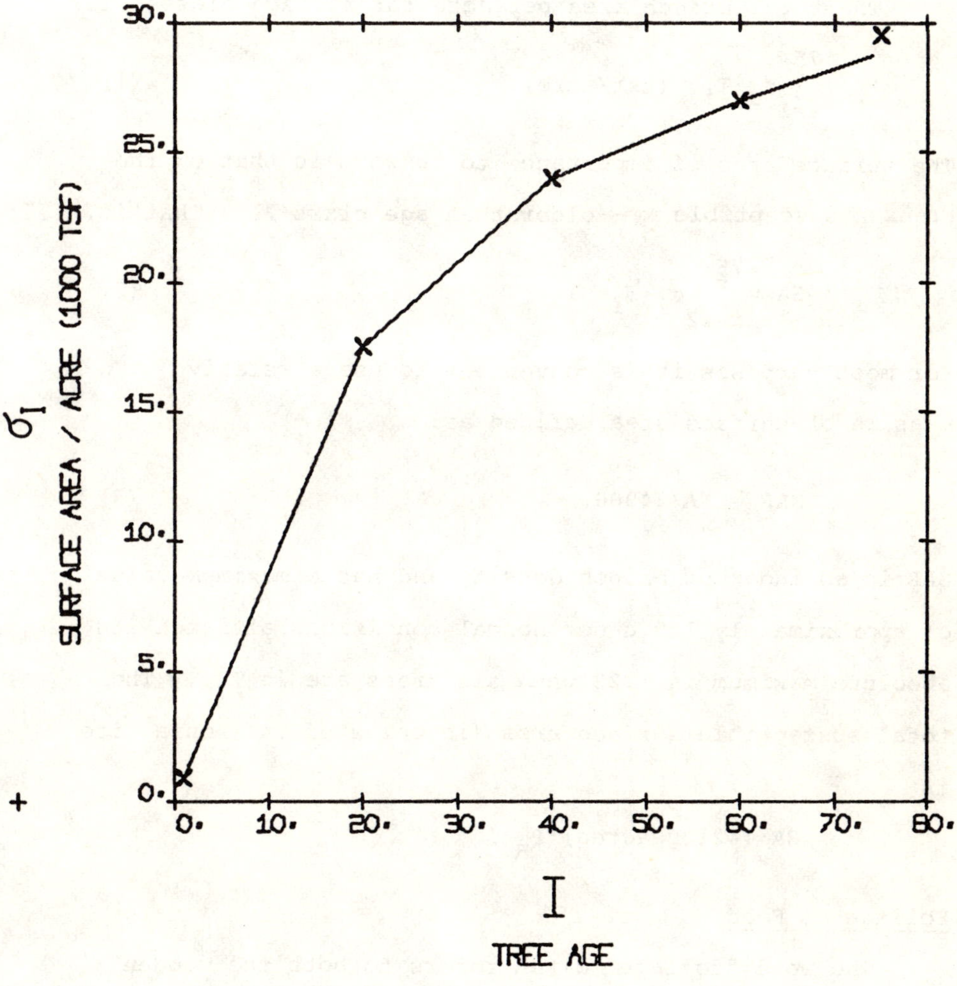

FIGURE 5: Brach Surface Area. The contribution of surface area in ten square feet/acre as a function of tree age.

The total branch area per acre for all age classes is:

$$\sum_{i=1}^{75} \sigma_i \cdot T_i \quad \text{(tsf/acre)}. \qquad (1)$$

The surface area of importance to budworm is that on the trees of susceptible age--older than age class 21. That is,

$$SA = \sum_{i=22}^{75} \sigma_i \cdot T_i \quad . \qquad (2)$$

For most purposes it is convenient to use a relative measure of surface area defined as:

$$SAR = SA/24000. \qquad (3)$$

SAR is an index of branch density and has a maximum value of approximately 1.0 under normal conditions although its absolute maximum is 1.23 when all trees are in T_{75}. The total susceptible surface area (in units of tsf) on a site is

$$SA \cdot (42100 \text{ acres}) \cdot P_H \; .$$

Foliage as Food

The word "foliage" often refers to both the wooden branches and the green needles. In this model, foliage refers only to the green needles. We assume that each unit of branch surface area SA has the same quantity of needles regardless of the age of the tree. Foliage is an average value representing conditions on the whole site.

Balsam fir retains its needles approximately eight years. To model all eight year classes would be an unnecessary computational burden. Because the budworm specifically prefers the new, first year foliage, it is necessary to distinguish this class from the older foliage. We therefore use two components of foliage:

F_1 = current, new foliage/tsf

F_2 = all older foliage/tsf

and $F_T = F_1 + F_2$ = total foliage. (4)

Rather than deal with a particular physical aspect of foliage (i.e., weight, volume, calories, etc.), we define an artificial foliage measure--the foliage unit (f.u.). A foliage unit is that quantity of new foliage found on ten square feet (tsf) of surface area when no budworm are present. That is, without budworm, $F_1 = 1.0$ f.u.

New foliage constitutes 26%, by weight, of the total foliage (Baskerville, pers. comm.). Therefore the nominal foliage levels are:

$F_1 = 1.0$ f.u.

$F_2 = 2.8$ f.u.

$F_T = 3.8$ f.u. (5)

Large larval feeding removes a quantity DEF_1 and DEF_2 of new and old foliage, respectively. Total defoliation is $DEF = DEF_1 + DEF_2$. The remaining foliage is F_i^*

$$F_1^* = F_1 - DEF_1$$

$$F_2^* = F_2 - DEF_2$$

$$F_T^* = F_1^* + F_2^* \tag{6}$$

These foliage levels remain until "winter." At that point tree mortality is applied (see below) and surviving trees are aged one year. Foliage is an average for all trees and must be compensated for the loss of the severely defoliated trees that have died. We assume that the foliage levels are lower on the trees that died than the ones that lived. If F_i^* is the average for living and dying trees, then

$$F_i^* = F_i^{**} \cdot R_L + F_i^{\,d} \cdot (1-R_L), \tag{7}$$

where F_i^{**} = Foliage level on trees that survive

$F_i^{\,d}$ = Foliage level on trees that die

R_L = Fraction of susceptible trees that live.

Next, we assume,

$$F_i^{\,d} = \delta \cdot F_i^{**}, \tag{8}$$

where δ is a constant. Its working value is taken as 0.5. Thus,

$$F_i^{**} = F_i^*/\{\delta\cdot(1-R_L)+R_L\} \ . \tag{9}$$

(Strictly speaking, R_L could be weighted by the surface area of the surviving trees.)

Foliage levels at the end of "winter" are F_1^{**} and F_2^{**}. At this time the previous year's new foliage becomes old foliage, some old foliage is lost, and a crop of new foliage is generated. A schematic diagram is in Figure 6. This process can be modeled in the framework of "survivals." First, the photosynthetic assimilation rate, A, is assumed proportional to the total foliage. For A scaled between 0 and 1.0 we have

$$A = \text{const.}\ F_T^{**} = \frac{1}{3.8}\cdot F_T^{**} \ . \tag{10}$$

The amount of new foliage generated is proportional to A:

$$F_0 = A \ . \tag{11}$$

The fraction of F_1^{**} that survives to old foliage is

$$S_{F_1} = 1.0 - 0.04\cdot F_1^{**} \ . \tag{12}$$

F_2^{**} survival is (see Figure 7)

$$S_{F_2} = (3.17 - 2.51\cdot A)\cdot A \ . \tag{13}$$

The crop of old foliage for year t+1 is

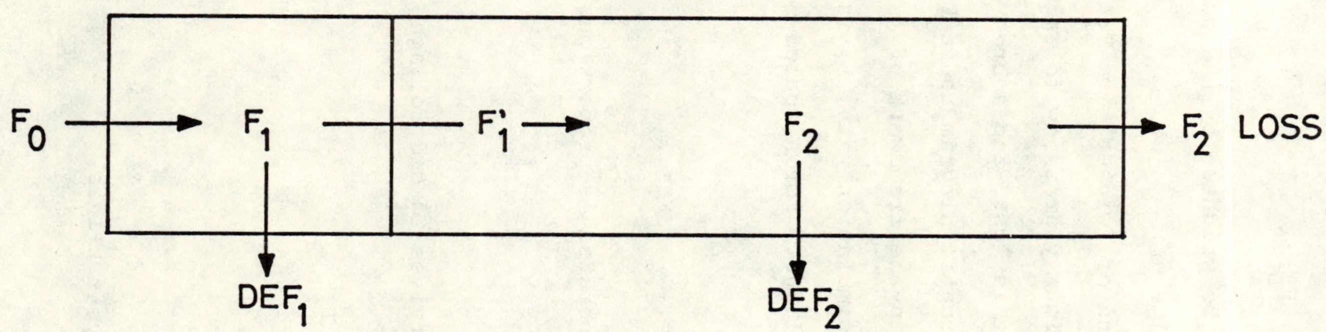

FIGURE 6: Schematic view of foliage. Fresh spring growth (F_0) replaces last year's new foliage (F_1) that has been partially eaten (DEF_1), the remainder becoming old foliage (F_2). Some old foliage is eaten (DEF_2) and some is lost to ageing (F_2 loss).

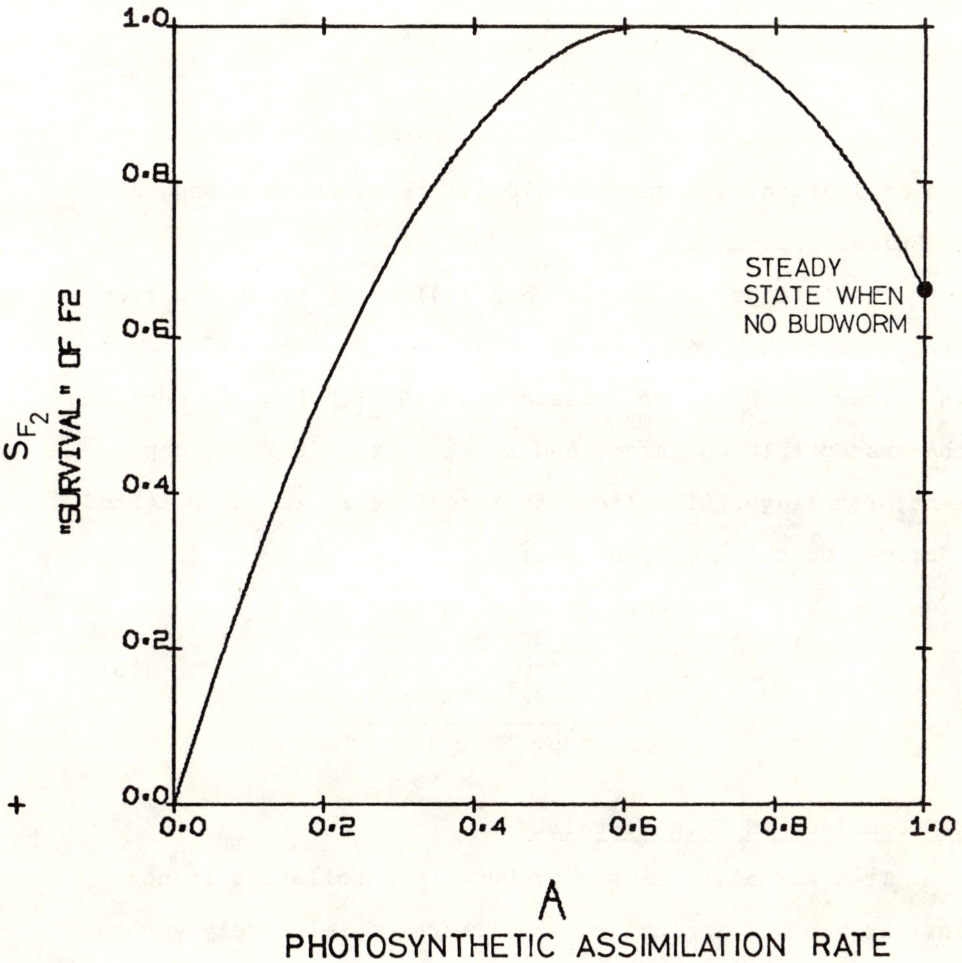

FIGURE 7: Survival of old foliage. The fraction of the old foliage after defoliation (F_2^*) remaining on the branch.

$$F_2' = S_{F_1} \cdot F_1^{**} + S_{F_2} \cdot F_2^{**} \tag{14}$$

The derivation of equations (10-14) is given in a separate technical appendix.

The foliage levels (11) and (14) apply to the surface area recovering from past defoliation. Additionally, there is a fresh quantity of surface area, SA_{21}, which is entering the susceptible age range and which carries a full complement of foliage. The final foliage levels are the weighted averages of this contribution:

$$F_1(t+1) = \frac{SA_{21} \cdot (1.0) + SA \cdot (F_0)}{SA_{21} + SA}$$

$$F_2(t+1) = \frac{SA_{21} \cdot (2.8) + SA \cdot (F_2')}{SA_{21} + SA} \tag{15}$$

Budworm Induced Tree Mortality

Tree mortality caused by budworm defoliation is not instantaneous. Typically, four years of relatively severe defoliation are required for noticeable levels of tree mortality. Severe defoliation means removal of nearly all new foliage. In reality, a complex of many physiological factors leads to the weakening of trees and their subsequent demise. In our model we assume that the "true" tree stress is correlated with the remaining foliage stock. Therefore, we use F_2^* as an index of physiological state and define stress, S_{BW}, as a function of this index. Mortality for any stress level is not the same for all age classes. We define an age specific susceptibility factor, μ_i, that reduces the effective mortality on the younger age classes. The mortality rate applied to age class i is then

$$M_i = S_{BW} \cdot \mu_I \tag{16}$$

where

$$S_{BW} = S_0 \cdot (1-F_2^*/2.8)^2 \; . \tag{17}$$

Equation 17 is illustrated in Figure 8. An iterative investigation of this function in the context of the site model indicated that S_0, the maximum stress level, should have a value of about 0.75. The susceptibility factor, μ_i, is shown in Figure 9. Susceptibility drops linearly for less mature trees; trees of intermediate age are half as susceptible as the oldest age class. Trees less than 22 years are not susceptible at all as they do not allow successful overwintering of small larvae.

As mentioned above, tree acreage that is removed because of budworm-induced mortality is replaced as age class one. The total amount of regeneration is

$$RG_{BW} = \text{Regeneration following budworm-induced mortality}$$

$$= \sum_{i=22}^{75} T_i \cdot M_i \; . \tag{18}$$

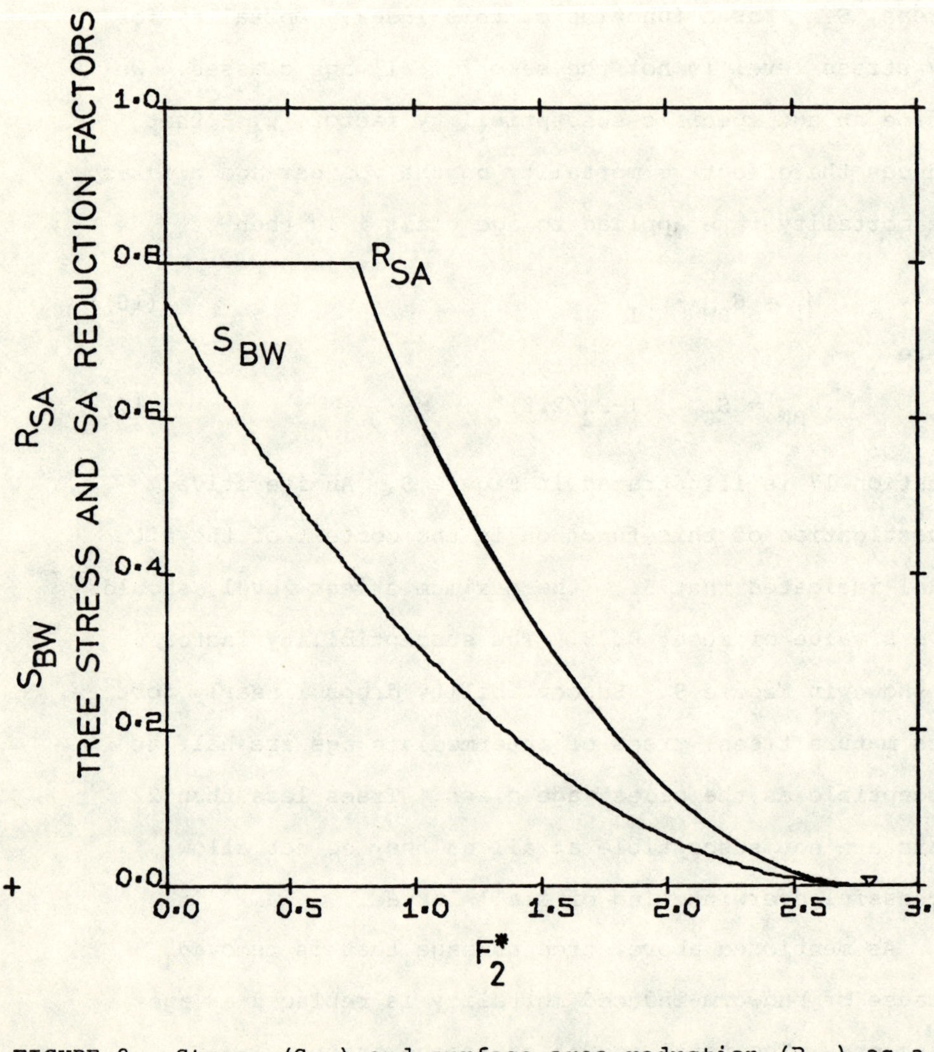

FIGURE 8: Stress (S_{BW}) and surface area reduction (R_{SA}) as a function of the amount of old foliage remaining.

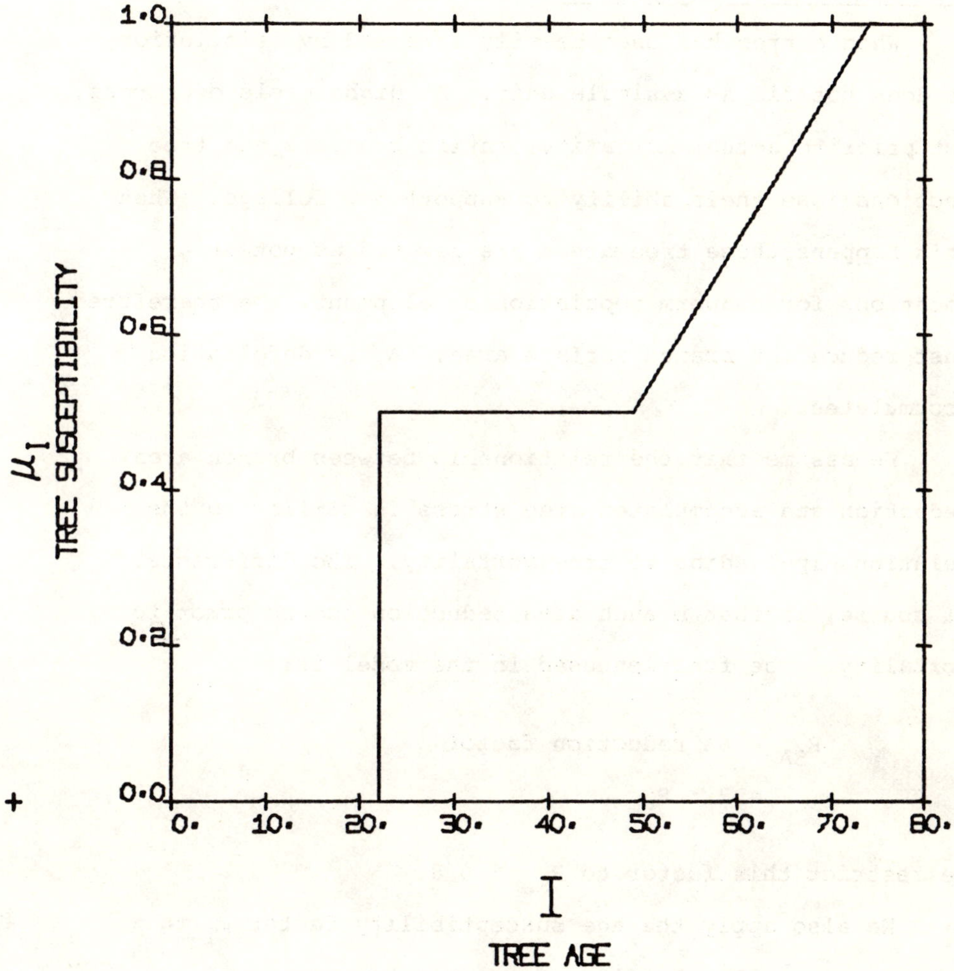

FIGURE 9: Age specific susceptibility. Susceptibility is relative to the oldest age class. Trees less than 22 years old are considered non-susceptible since small larvae have very low over-wintering success on them.

Branch Surface Area Reduction

When a tree has been heavily stressed by defoliation it does not die as a single unit. At high levels of stress, but prior to actual mortality, entire branches and tree sections lose their ability to support new foliage. When this happens those tree areas are removed as potential locations for budworm population development. We therefore must reduce the branch surface area, SA, as defoliation accumulates.

We assume that the relationship between branch area reduction and accumulated tree stress is similar to the relationship leading to tree mortality. The difference, of course, is that branch area reduction occurs prior to mortality. The function used in the model is:

$$R_{SA} = \text{SA reduction factor}$$
$$= 2 \cdot S_{BW} \; . \qquad (19)$$

We restrict this factor to $R_{SA} \leq 0.8$.

We also apply the age susceptibility factor μ_i to surface area. The total surface area becomes

$$SA = \sum_{i=22}^{75} \sigma_i \cdot T_i \cdot (1 - \mu_i \cdot R_{SA}) \; . \qquad (20)$$

Tree mortality and branch surface area loss occur continuously throughout the year. However, for convenience in our model we consolidate these processes and update the tree age classes at one specific time each year. We choose that time as "winter"; during the small larval stage. This sequence implicitly assumes that females cannot distinguish oviposition sites that are "healthy" from those that are about to be removed through tree mortality. Thus, eggs are laid evenly on all foliage and when surface area is lost egg density does not change, though the total of eggs per site is reduced.

BUDWORM SURVIVAL

Introduction

This section describes how the life cycle of the spruce budworm has been modeled. The style is principally descriptive; explicit enumeration of the equations used can be found in Appendix A. Details on the rationale for developing the present forms are given in a separate technical appendix.

To a major extent the formulation used here parallels that given in the budworm monograph (Morris, 1963). Principal life stages are modeled in such a way as to reflect the main influences described in that reference. Additionally, the survival level of each life stage in the model has been scaled independently so that its characteristic magnitude is approximately that given in the monograph. For example, during the outbreak in the late 1950s pupal survival was found to average 0.66. The parameters of pupal survival in the model have been adjusted so that during outbreaks it too approximately averages this value.

As is illustrated in Figure 2 the budworm cycle can be started at any convenient point. We shall find it useful to begin our descriptions with the large larval stage, that is Instars III-VI. During this stage the majority of significant processes occurs and this particular life stage also contributes critically to the overall dynamics of the budworm/forest system. This stage produces the defoliation that affects the forest itself. Feeding levels here set the nutritional conditions that control many of the later processes through the adult and egg-laying periods. It is the large larvae that are subject to the important bird predation. Also, this stage is the target for insecticide spraying.

Morris (1963) found that the large larval stage was statistically the most important with respect to variation in population levels. A functional analysis shows that this stage also governs the dynamic behaviour of the budworm/forest system.

Feeding by Large Larvae

Budworm preferentially consume new foliage, but they will consume old foliage if populations are high. The feeding response functions used in the model incorporate the competitive saturation effect of large budworm numbers on a finite food resource. We let d_0 be the maximum foliage consumption rate for an individual large larva during the feeding season. This value is set at $d_0 = 0.0074$ foliage units consumed per season. Note that at this feeding rate a budworm density of 135 larvae/tsf would remove all the new foliage, and a density of 513 larvae/tsf would remove all the new plus all the old foliage. This rate is not realistic at high population densities because of the competition between individuals. We therefore define a value d_1 as the actual amount of new foliage eaten per individual per season.

A standard competition function is used:

$$d_1 = \frac{F_1}{N_L}\left(1-\exp\left(-\frac{d_0 \cdot N_L}{F_1}\right)\right) \tag{21}$$

where N_L is the density of large larvae per tsf. The relative consumption rate is $DR1 = d_1/d_0$. This function is illustrated in Figure 10 for ten levels of F_1.

The difference between the actual new foliage consumption, d_1, and the maximum desired, d_0, is the

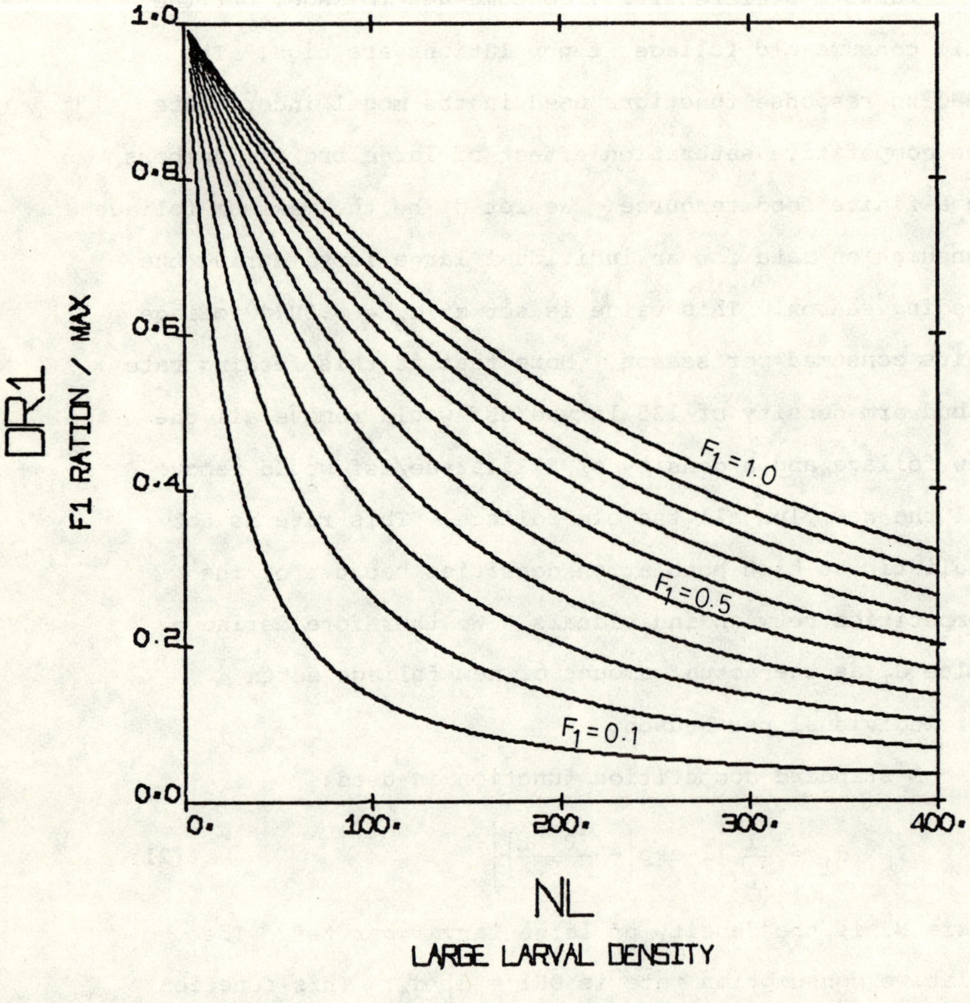

FIGURE 10: New foliage ration. Amount of new foliage eaten per larva relative to the maximum possible as a function of larval density and available new foliage.

residual "demand" for old foliage. Larvae attempt to satisfy this demand by feeding on old foliage through a function analogous to equation (21). The total consumption (relative to the maximum d_0) is

$$DR = DR1 + DR2.$$

This function is illustrated in Figure 11 for ten levels of F_T.

The consumption of each foliage type times the large larval density is the amount of that foliage type removed per ten square feet of branch surface area. This consumed foliage is subtracted from the foliage stock. Foliage levels in the defoliated state are indicated by the superscript "*" (e.g., $F_1^* = F_1 - DEF_1$).

Survival of Large Larvae

We consider four major processes affecting the survival of large larvae:

(1) food limitation

(2) weather

  (i) through the intermediary of disease

  (ii) through other related factors

(3) parasitism

(4) predation.

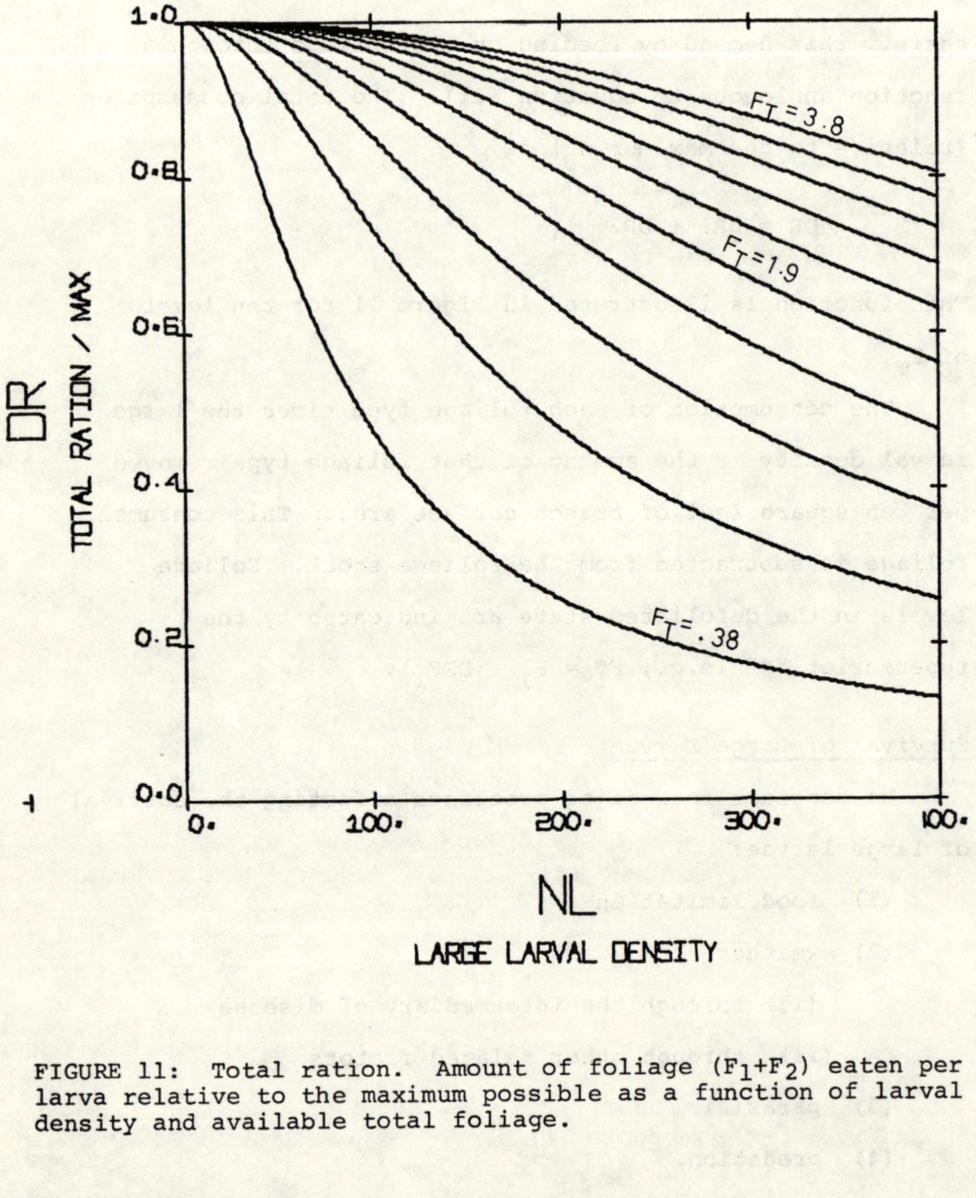

FIGURE 11: Total ration. Amount of foliage (F_1+F_2) eaten per larva relative to the maximum possible as a function of larval density and available total foliage.

First, the effects of food limitation are modeled by assuming that survival is proportional to the average amount of food consumed. Second, the effects of weather are assumed to affect overall survival through an aggregate weather factor, W_F. We use a three level weather index: 1, 2, and 3. Weather 1 represents generally cool and damp summers; Weather 3 represents warm and dry. Weather class 2 is the long-term average; its weather factor is 1.0. Survival factors for other weathers are relative to this average value. When the weather class is 1, the weather factor is 0.76; for class 3 it is 1.29.

The rate of parasitism is a decreasing function of larval density. The maximum value is 40% at low densities and decreases exponentially such that at 230 larvae/tsf parasitism is only 20%.

$$\text{Parasitism} = 0.4\, e^{-.003 \cdot N_L} .$$

The survival from parasitism is thus

$$\begin{aligned} S_L &= \text{Survival from parasites} \\ &= 1 - 0.4\, e^{-.003 \cdot N_L} . \end{aligned} \tag{22}$$

The combined survival leading to sixth instar larvae is

$$\begin{aligned} S_L'' &= \text{Survival to Instar VI} \\ &= k_L \cdot S_L' \cdot DR \cdot W_F . \end{aligned} \tag{23}$$

The scaling parameter k_L is set at 0.425. This function is illustrated in Figure 12 for various levels of initial foliage. Figure 13 shows the range of this survival for each weather class when $F_T = 3.8$.

The derivation of the predation function has been covered extensively elsewhere (Holling, 1976). Predation is limited to birds feeding on sixth instar larvae. The net survival for large larvae is

$$S_L = S_L'' \cdot S_{PRED} \; . \tag{24}$$

This survival is shown in Figure 14. This figure differs from Figure 12 only by addition of predation.

The density of pupae resulting from these processes is given by

$$N_P = S_{PRED} \cdot N_6 = S_L \cdot N_L \; . \tag{25}$$

Insecticide spraying of large larvae reduces the density of N_6 larvae but does not otherwise change the rate of survival S_L.

During the large larval stage many mortality factors are set in motion that are not realized until the budworm are pupae. This effect correlates the survival of pupae with the survival of large largae. The function used is

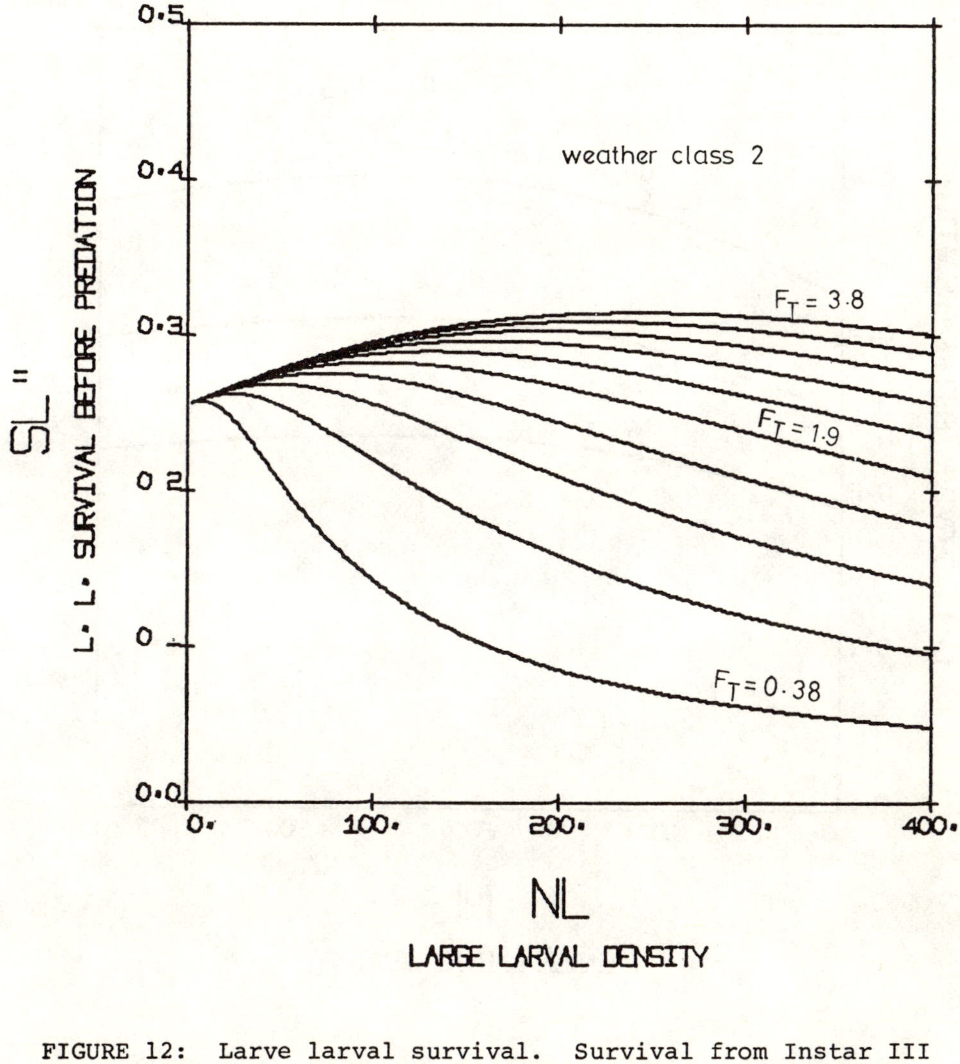

FIGURE 12: Larve larval survival. Survival from Instar III to VI (before predation) for different amounts of available foliage and average weather (2).

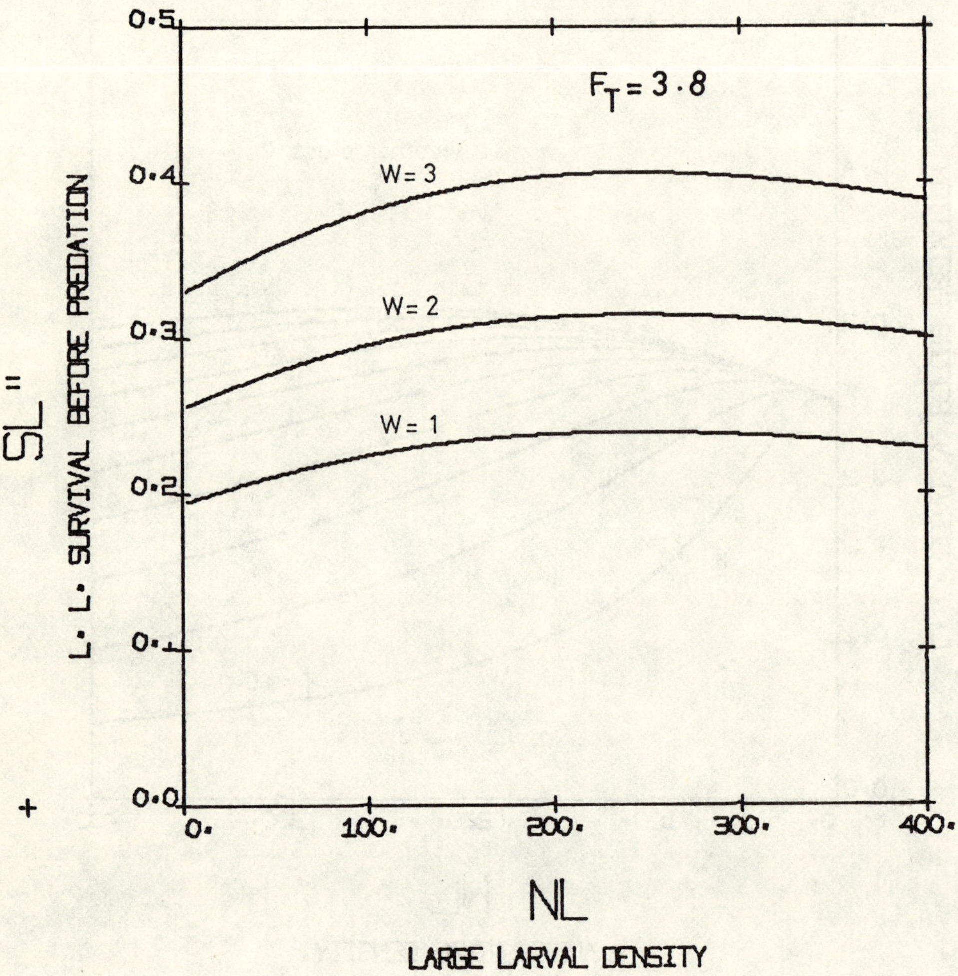

FIGURE 13: Weather. Relative levels of survival for each weather class when there is maximum total foliage.

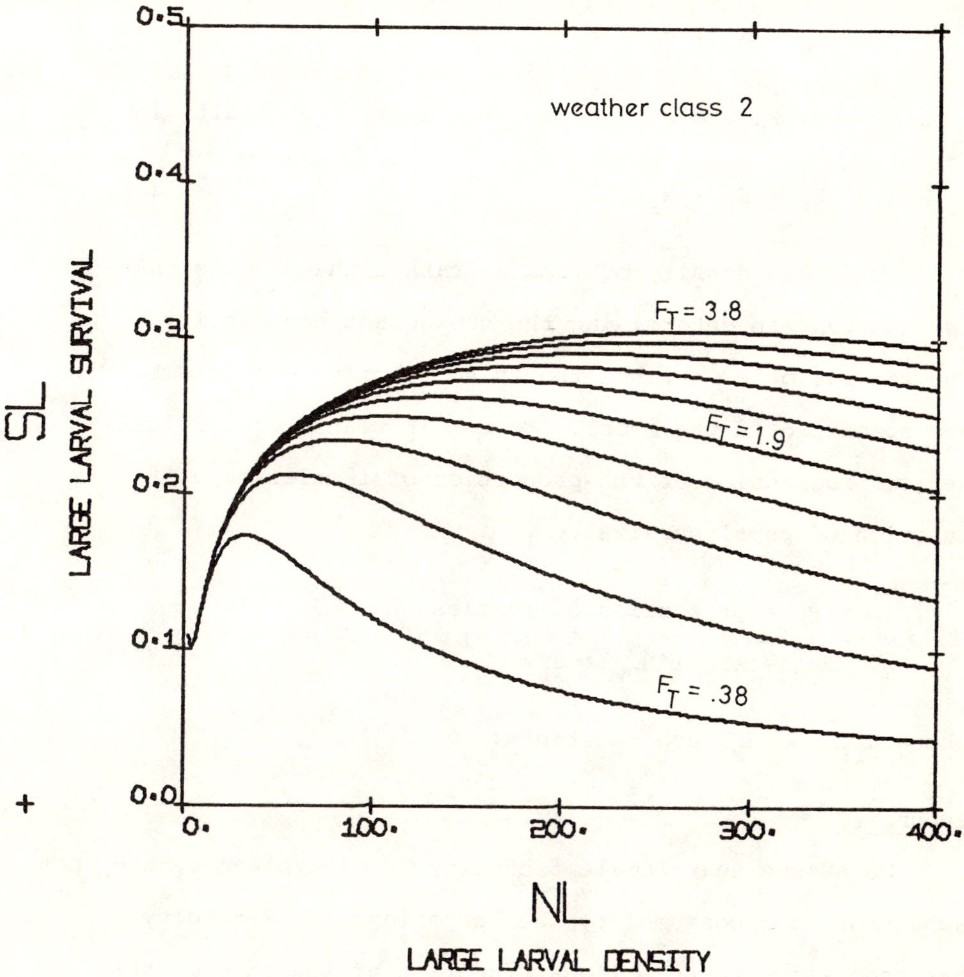

FIGURE 14: Large larval survival. Survival of large larvae from all factors except insecticide spraying for average weather and various levels of available foliage.

$$S_P = A_P + B_P \cdot S_L \tag{26}$$

where A_P and B_P are constants. The density of adults is

$$N_A = S_P \cdot N_P \ . \tag{27}$$

It is the density of females rather than adults that is important in determining the subsequent egg density. The sex ratio is variable though no specific mechanisms are specified in the model. We use instead a statistically derived regression of the proportion of females as a function of pupal survival:

$$P_F = \text{proportion of females}$$
$$= A_{PF} + B_{PF} \cdot S_P \ , \tag{28}$$

where A_{PF} and B_{PF} are constants.

Fecundity

We assume that female fecundity is determined by the amount of food consumed by the large larvae. Fecundity has been shown to be a linear function of the size of the pupal case. We retain this relationship and assume that pupal size is proportional to the cube root of larval weight. We do not know the conversion efficiency of foliage into larval weight but assume it to be linear with food consumption. The weight versus foliage consumption equation is

$$\text{WGT} = \text{weight of pupae (relative units)}$$

$$= A_{F1} \cdot DR1 + A_{F2} \cdot DR2 + B_F \qquad (29)$$

where $A_{F1} = 34.1$, $A_{F2} = 24.9$, $B_F = -3.4$.

The fecundity regression equation in terms of pupal case size, x, is

$$\text{FEC} = \text{fecundity}$$

$$= 165.64 \cdot x - 328.52.$$

We substitute the cube root of pupal weight to get

$$\text{FEC} = 165.64\ (\text{WGT})^{1/3} - 328.52. \qquad (30)$$

This equation is shown in Figure 15 for various levels of initial foliage. Fecundity is restricted to be greater than 40 eggs per female. If nutrition were so low as to produce fewer than 40 eggs, the pupae would not have survived.

At this point we have a potential density of eggs generated by the population of budworm on the site. This egg density is

$$e_g = \text{FEC} \cdot S_F \cdot N_F = \text{FEC} \cdot P_F \cdot S_F \cdot N_A \qquad (31)$$

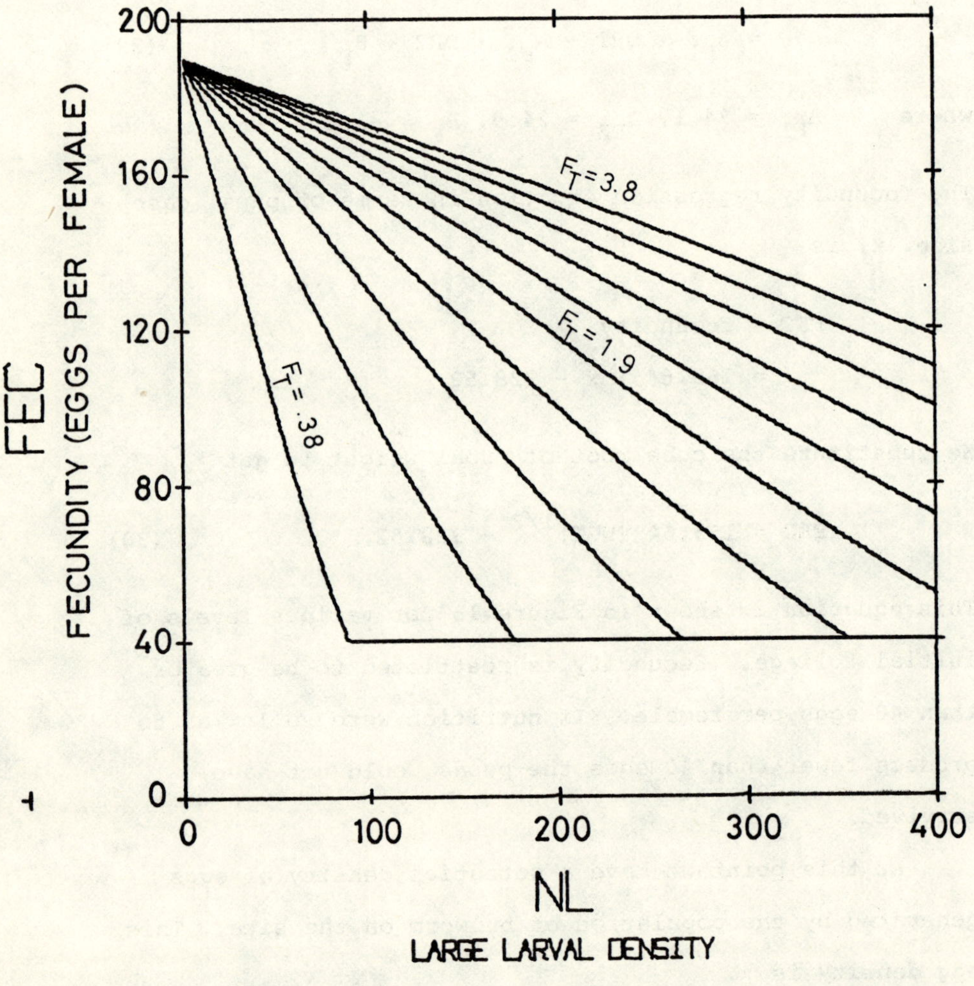

FIGURE 15: Fecundity. The number of eggs carried per female as a function of original larval density and available foliage.

where S_F is the survival of females (equal to 1.0). This egg density is plotted against initial larval density for several values of foliage in Figure 16.

Egg Dispersal

It is convenient to think of eggs migrating rather than adults, as an individual female will deposit some of her eggs on the site and remove some to other locations. The number remaining on the site depends upon the mobility of the female that is in turn governed by the weight of egg mass she is carrying. Thus when fecundity is high the fraction of eggs migrating is low. The form used is given as

$$F_{MIG} = \text{fraction of eggs migrating off site}$$
$$= 1.1 = FEC/200$$

with the constraints:

$$0.35 \le F_{MIG} \le 0.90. \qquad (32)$$

The form of this function is shown in Figure 17. This function separates the eggs generated, e_g, into those that stay on the site and those that migrate off.

When the site model is used as a stand-alone entity the dispersal characteristics of the full simulation can only be approximated. The discrepancy between the

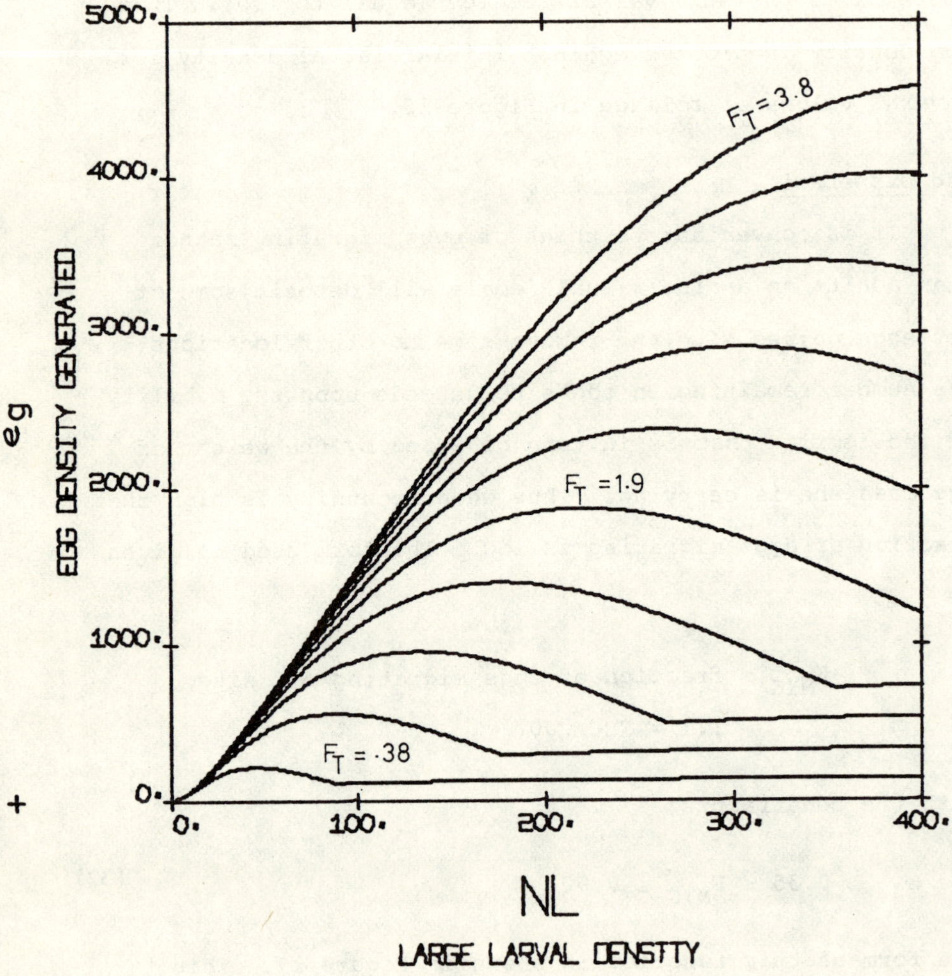

FIGURE 16: Generated egg density. The total number of eggs generated on site as a function of original larval density and available foliage. Some of these eggs will remain on site; the remainder will disperse away.

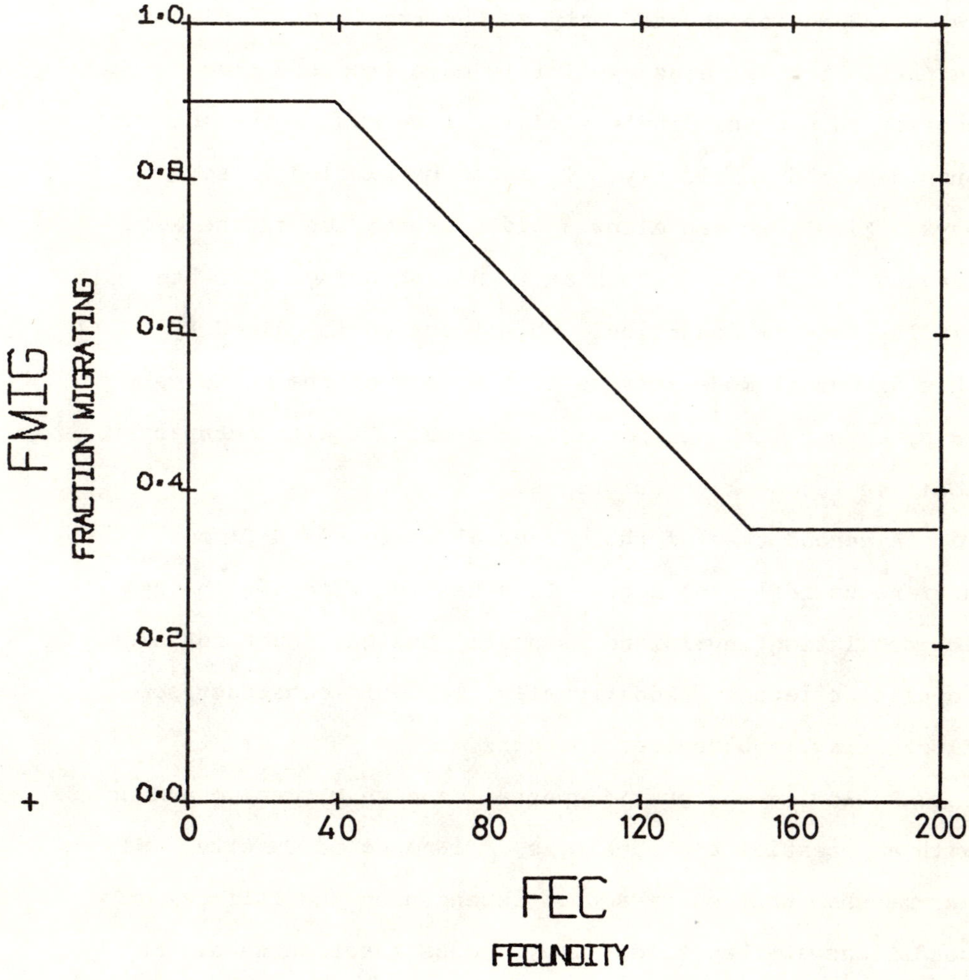

FIGURE 17: The fraction of eggs migrating off site is a function of the number of eggs produced.

two model scales depends only on the treatment of dispersal. In the full model eggs explicitly move from one site to another, but in the single-site model we must represent the outside world implicitly. This can be handled in several ways. First, we can allow a given proportion of the eggs leaving the site to return as if neighbouring sites had similar budworm densities. This fraction is called R_{in}. This dispersal mode acts as if the rest of the world were in synchrony with our particular site. It also acts as a positive feedback to outbreaks.

A second type of the dispersal is to set a fixed extraneous influx of eggs. This has the effect of boosting low population levels and is nearly insignificant for high population levels. Additionally, we could construct some fluctuating input pattern of eggs.

Following whatever dispersal mode we choose, we end up with a potential egg density, e_p, landing on the site. We assume that this success will depend upon the foliage conditions at that time. The success function we use is strictly descriptive. Its form is

$$P_1 = \left(\frac{F_T^*}{3.8}\right) \cdot \left\{2.0 - \left(\frac{F_T^*}{3.8}\right)\right\} , \tag{33}$$

and is illustrated in Figure 18.

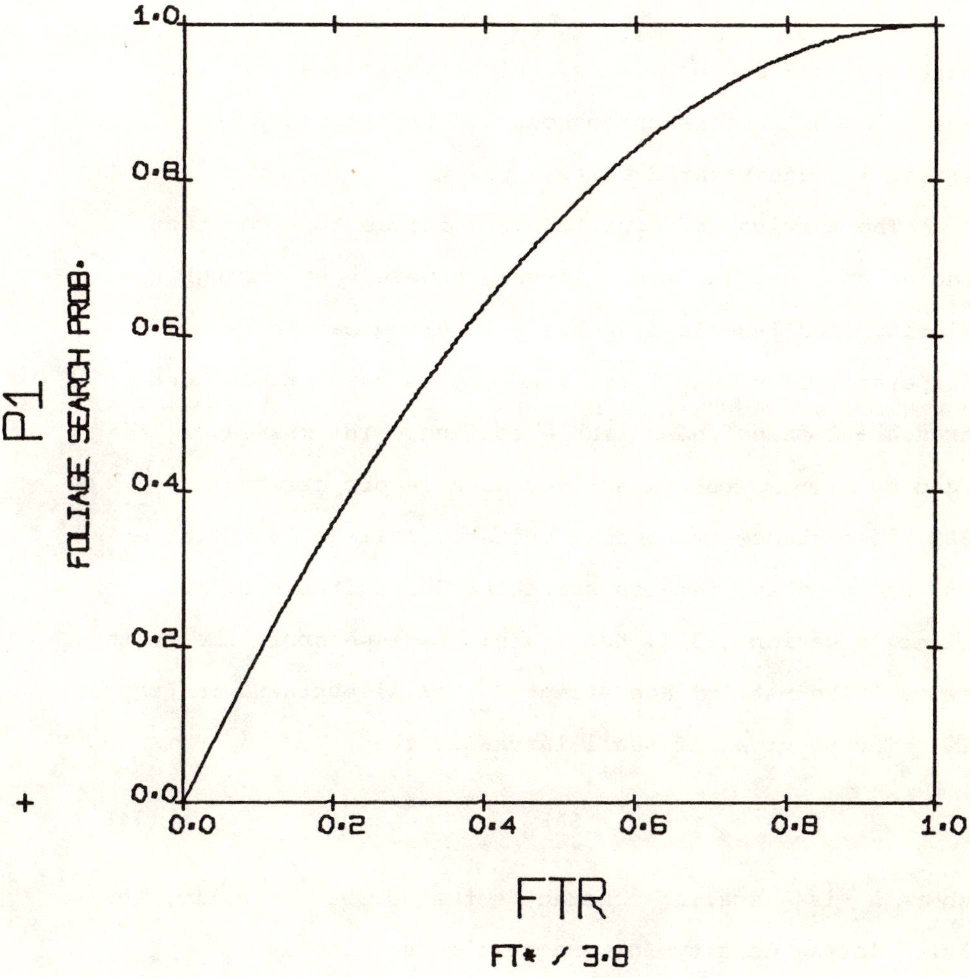

FIGURE 18: The probability that a female will find an adequate oviposition site depends on the amount of available foliage.

At this point in the simulation sequence forest mortality and ageing occurs. Although in reality this is a continuous process throughout the entire year, its effect is concentrated at this point.

The survival of eggs to small larvae is a constant factor of 81%. The small larvae, however, go through two distinct small-scale dispersals. During each of these dispersals they must first land upon a susceptible host branch and then find suitable foliage. The chance of landing on an appropriate surface area is proportional to SAR. The chance of finding suitable foliage is the same as that used for females searching for suitable oviposition sites [equation (33)], but in this case we apply the factor twice. The net foliage effect P_{ss} is illustrated in Figure 19. The survival of small larvae is then

$$S_s = k_s \cdot P_{ss} \cdot SAR \tag{34}$$

where k_s is a scaling constant set at 0.352. Finally, the large larvae density for the next generation is

$$N_L(t+1) = S_s \cdot N_s \ . \tag{35}$$

The ratio of large larvae in one generation to that of the previous generation is the growth rate, or recruitment factor. This function for various levels of surface

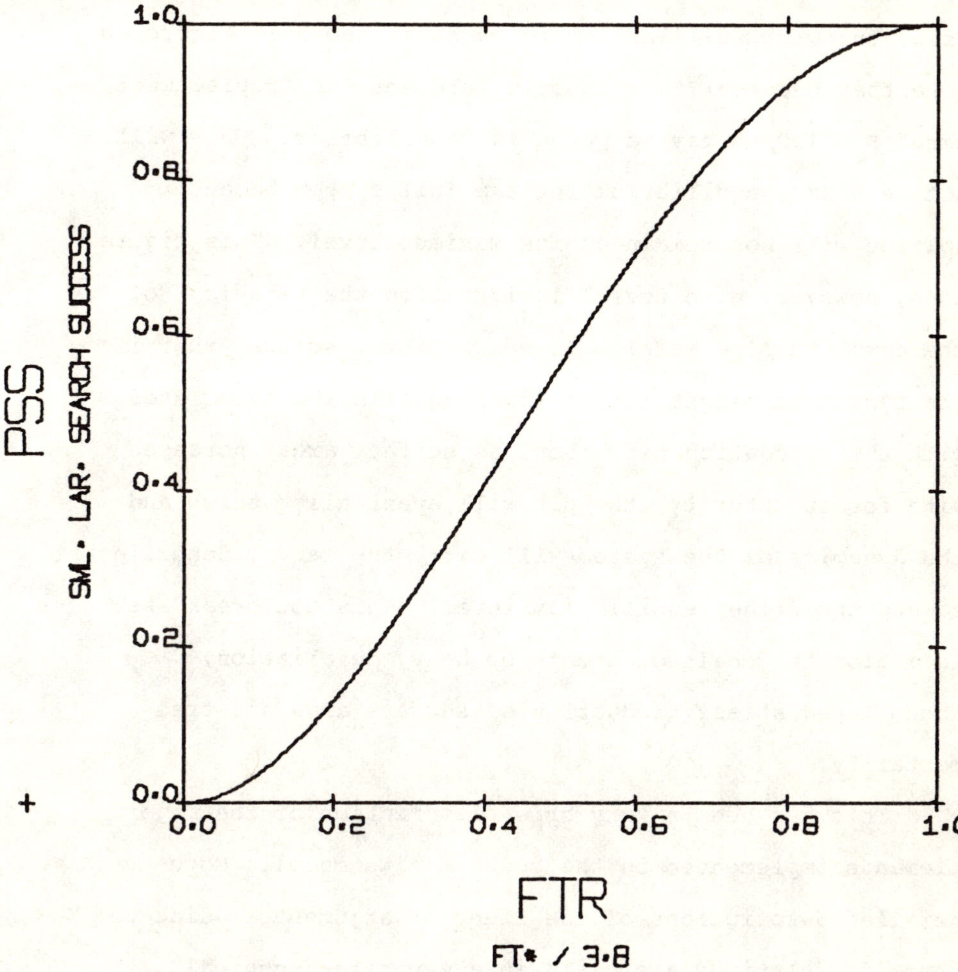

FIGURE 19: The net probability that the small larvae will find adequate foliage during their two dispersals depends on the amount of available foliage.

area, but with full initial foliage, is shown in Figure 20. Note that whenever this function crosses the "replacement line" R = 1.0, we have a potential equilibrium. This will not be a true equilibrium for the full system because foliage will not remain at its maximum level. This figure does, however, give useful insight into the behaviour of the complete site model. It shows, for instance, that if the system is caught at the lower equilibrium associated with the "predation pit," then, as surface area increases with forest maturity, the pit will eventually vanish and the dynamics of the system will carry the larval densities toward the higher equilibrium level. This "outbreak" sets in motion the chain of events of heavy defoliation, accumulated stress, reduction of surface area, and tree mortality.

This has been a very brief description of the major elements implemented in the budworm site model. More detailed descriptions of the lines of argument leading to these formulations are given in a separate technical appendix. A complete set of the actual equations is given in Appendix A. A thorough examination of the implications and assumptions implicit in this formulation and implicit in those things left out has not been given. Many of these are recognized and accepted under the banner

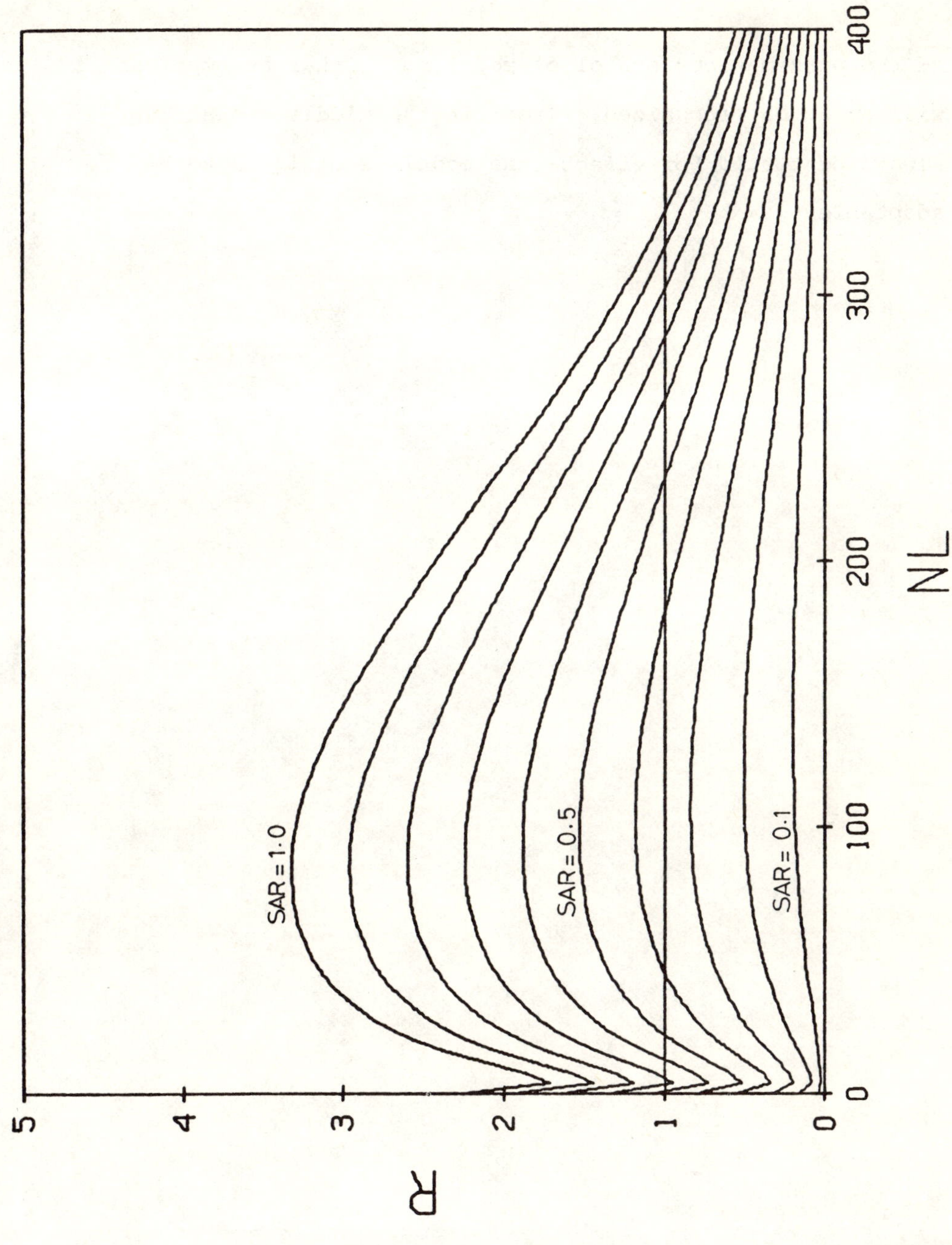

FIGURE 20: Growth rate. This is the ratio of N_L next year over N_L this year as a function of this year's density. The curves are for different levels of the branch density index (SAR) and for F_T = 3.8.

of the parsimonious school of modeling. Other assumptions will never be recognized. Those in the middle ground can always be tested for effect--the model is designed to be adaptable.

REFERENCES

Holling, C. S. 1976. Dealing with inadequate data. (In preparation.)

Holling, C. S., G. B. Dantzig, G. Baskerville, D. D. Jones, and W. C. Clark and R. M. Peterman 1977. Quantitative evaluation of pest management options: the spruce budworm case study. In: *Proc. Symp. On Concepts and Practice of Integrated Pest Management in Forestry*, W. F. Waters (ed). (In press.)

Holling, C. S., D. D. Jones, and W. C. Clark. 1976. The spruce budworm management study: fables of ecological policy design. (This volume.)

Morris, R. F. 1955. The development of sampling techniques for forest insect defoliators, with particular reference to the spruce budworm. *Can. J. Zool.* *33*:225-294.

Morris, R. F. (ed.) 1963. The dynamics of epidemic spruce budworm populations. Mem. Ent. Soc. of Canada, Vol. 31.

APPENDIX A - SUMMARY OF EQUATIONS

This appendix summarizes the equations used in the Budworm Site Model. Variables and parameters are minimally defined; for explanation and derivation consult the text and the separate technical appendix. Where appropriate, equations are referenced to text figures. The symbols used here are paired with their FORTRAN equivalents in the glossary, Appendix B.

1. <u>Feeding by Large Larvae</u>

d_0 = maximum consumption rate/individual/season

= 0.0074 f.u.

d_1 = new foliage eaten/individual/season

DR1 = d_1/d_0 = relative consumption of F_1

$$= \frac{F_1}{d_0 N_L}\left(1 - \exp -\left(\frac{d_0 \cdot N_L}{F_1}\right)\right) \qquad \text{(A.1)}$$

(Fig. 10)

where

N_L = density of large larvae/tsf

$d_0 - d_1$ = individual "demand" for old foliage

DR2 = relative consumption of F_2

$$= \frac{F_2}{d_0 \cdot N_L}\left[1 - \exp -\left(\frac{(d_0-d_1)N_L}{F_2}\right)\right] \qquad \text{(A.2)}$$

NOTE: If the argument of the exponent becomes too small (<.001, say), we use the expansion:

$$(1-e^{-z}) = z(1 - \frac{z}{2} + \frac{z^2}{6} \ldots) \qquad \text{(A.3)}$$

DR = total consumption relative to d_0

= DR1 + DR2 (Fig. 11)

2. Removal of Foliage

DEF_1 = total F_1 removed/tsf

$$= d_1 \cdot N_L \qquad \text{(A.4)}$$

$$DEF_2 = d_2 \cdot N_L = DR2 \cdot d_0 \cdot N_L$$

$$\left.\begin{aligned} F_1^* &= F_1 - DEF_1 \\ F_2^* &= F_2 - DEF_2 \\ F_T^* &= F_1^* + F_2^* \end{aligned}\right\} \quad F_i^* \rightarrow \text{foliage in the defoliated state} \qquad \text{(A.5)}$$

3. Survival of Large Larvae

Parasitism = $0.4\ e^{-.003 \cdot N_L}$

thus (A.6)

$$S_L' = 1 - 0.4\ e^{-.003 \cdot N_L}$$

= survival from parasites

S''_L = survival to Instar VI

$$= k_L \cdot S'_L \cdot DR \cdot W_F \tag{A.7}$$

where

W_F = weather factor

class	W_F
1	0.76
2	1.00
3	1.29

k_L = constant = 0.425

The density of Instar VI is

$$N_6 = S''_L \cdot N_L \tag{A.8}$$

Predation is caused by three classes of vertebrate (bird) predators.

Prey = effective density of Instar VI per tsf

$$= N_6 \cdot \left(\frac{3.8}{F_T}\right) \tag{A.9}$$

$Pred_i$ = predation rate from predator type i

$$= \frac{A_i}{e^{-\alpha_i Prey} + B_i \cdot Prey} \tag{A.10}$$

where

A_i, B_i and α_i are the predation coefficients for predator type i. The total (instantaneous) functional response is

$$PRED = \sum_{i=1}^{3} Pred_i$$

$$S_{PRED} = \left(1 + \frac{PRED}{2 \cdot SAR \cdot AK}\right)^{-AK} \tag{A.11}$$

= survival from predation

Total survival of large larvae is

$$S_L = S''_L \cdot S_{PRED} \qquad (A.12)$$

(Fig. 14)

Pupal density is

$$N_p = S_{PRED} \cdot N_6 = S_L \cdot N_L \qquad (A.13)$$

4. Survival of Pupae

$$S_P = A_P + B_P \cdot S_L \qquad (A.14)$$

where

$A_P = 0.473$

$B_P = 0.828$

$$N_A = S_P \cdot N_P = \text{density of adults/tsf} \qquad (A.15)$$

5. Fecundity

WGT = weight of pupae (relative units)

$$= A_{F1} \cdot DR1 + A_{F2} \cdot DR2 + B_F \qquad (A.16)$$

where

$A_{F1} = 34.1,\ A_{F2} = 24.9,\ B_F = -3.4$

FEC = fecundity

$$= 165.64(WGT)^{1/3} - 328.52 \qquad (A.17)$$

(Fig. 15)

Fecundity is constrained such that

$FEC \geqq 40$

6. Production of Eggs

P_F = proportion of females

$$= A_{PF} + B_{PF} \cdot S_P \qquad \text{(A.18)}$$

where

$A_{PF} = 0.289$

$B_{PF} = 0.237$

$$e_g = FEC \cdot P_F \cdot N_A \qquad \text{(A.19)}$$

= egg density generated on site (Fig. 16)

where

S_F = survival of females = 1.0

7. Migration of Eggs

F_{MIG} = fraction of eggs migrating off site

$$= 1.1 - FEC/200 \qquad \text{(A.20)}$$

with the constraints: (Fig. 17)

$0.35 \leq F_{MIG} \leq 0.90$

$e_s = e_g \cdot (1-F_{MIG})$ = eggs staying on site (A.21)

$$E_0 = e_g \cdot F_{MIG} \cdot P_H \cdot SAR \qquad \text{(A.22)}$$

= eggs leaving/acre (÷24,000)

$$E_i = R_{in} \cdot E_0 + E^* \qquad \text{(A.23)}$$

= in-dispersing eggs/acre (÷24,000)

R_{in} = ratio of E_0 returning

E^* = extraneous egg source

$$e_p = e_s + E_i \cdot T_{sp}/SAR \qquad \text{(A.24)}$$

= potential egg density on site

where

T_{sp} = fraction of trees that are of susceptible age (≧ 22 years)

$= \sum_{i=22}^{75} T_i$

$P_1 = (F_T^*/3.8)\{2.0 - (F_T^*/3.8)\}$ (A.25)

= probability of finding suitable foliage for oviposition (Fig. 18)

$N_E = e_p \cdot P_1$

= egg density/tsf (A.26)

NOTE: At this point Subroutine FOREST is called; see Sections 10-14 below.

8. Survival of Eggs

$S_E = 0.81$

$N_S = S_E \cdot N_E$ = density of small larvae (A.27)

9. Survival of Small Larvae

$P_2 = (F_T^{**}/3.8)\{2.0 - (F_T^{**}/3.8)\}$ (A.28)

$P_{ss} = P_1 \cdot P_2$ (A.29)

= probability of successful search for foliage during fall and spring dispersals (Fig. 19)

$S_S = k_s \cdot P_{ss} \cdot SAR$ (A.30)

where

$$k_S = \text{constant} = 0.352$$

$$N_L(t+1) = S_S \cdot N_S \qquad \text{(A.31)} \quad \text{(Fig. 20)}$$

10. Budworm-Induced Tree Mortality

$$F_R = F_2^*/2.8$$

$$S_{BW} = S_0 \cdot (1-F_R)^2 \qquad [S_o = 0.75] \qquad \text{(A.32)}$$

= budworm-induced stress (Fig. 8)

$$[S_{BW} = 0 \text{ if } F_R > 1.0]$$

$$R_{SA} = 2 \cdot S_{BW}$$

$$[R_{SA} \leqq 0.8] \qquad \text{(A.33)}$$

= reduction in surface area factor

11. Removal of Dead Trees

$$M_i = S_{BW} \cdot \mu_i = \text{mortality rate for } T_i \qquad \text{(A.34)}$$

(μ_i: cf. Section 16)

$$T_i^* = T_i \cdot (1-M_i) \qquad \text{(A.35)}$$

RG_{BW} = regeneration of land from budworm-induced mortality

$$= \sum_{i=22}^{75} T_i \cdot M_i \qquad \text{(A.36)}$$

"Natural" mortality is applied to T_{74} and T_{75} at the rate of 2.37% per annum.

$$T_i^{**} = T_i^*(1-.0237) \qquad (i = 74,75) \qquad \text{(A.37)}$$

$$RG_{75} = 0.237(T_{74}^* + T_{75}^*) \qquad \text{(A.38)}$$

12. Tree Vector Update (Fig. 4)

$$T_{75}(t+1) = T_{74}^{**}(t) + T_{75}^{**}(t)$$

$$T_{i=1}(t+1) = T_i^*(t) \qquad (i = 1, \ldots, 73) \qquad \text{(A.39)}$$

$$T_1(t+1) = RG_{BW} + RG_{75}$$

13. Surface Area, Susceptible Host and Matriculation

$$SA = \sum_{i=22}^{75} \sigma_i \cdot T_i \qquad (\sigma_i: \text{ cf. Section 16})$$

$$= \text{susceptible surface area} \qquad \text{(A.40)}$$

$$T_{sp} = \sum_{i=22}^{75} T_i = \text{proportion of host land of susceptible age} \qquad \text{(A.41)}$$

$$SAR = SA/24000$$

$$\overline{T_{21}} = \frac{1}{3}(T_{20}+T_{21}+T_{22}) = \text{measure of host land entering the susceptible population} \qquad \text{(A.42)}$$

$$SA_{21} = \overline{T_{21}} \cdot \sigma_{21} = \text{surface area becoming susceptible} \qquad \text{(A.43)}$$

14. Foliage Compensation for Dead Trees

The foliage levels remaining after feeding, F_i^*, are assumed to be an average of that on trees currently dying and that on trees surviving.

$$F_T^* = F_T{}^d \cdot R_D + F_T^{**} \cdot R_L$$

where

F_T^{**} = foliage level on trees that escape this year's mortality

$F_T{}^d$ = foliage on those that die

$= \delta \cdot F_T^{**}$ $(\delta = 0.5)$

R_L = fraction of T_{sp} that lives

$$= 1 - \frac{RG_{BW}}{T_{sp}^*} \qquad (A.44)$$

thus

$$F_T^* = F_T^{**} \cdot \{\delta(1-R_L) + R_L\}$$

The compensated foliage levels are:

$$F_T^{**} = F_T^*/\{\delta(1-R_L) + R_L\}$$
$$F_1^{**} = F_1^*/\{\delta(1-R_L) + R_L\} \qquad (A.45)$$
$$F_2^{**} = F_2^*/\{\delta(1-R_L) + R_L\}$$

NOTE: The value of T_{sp}^* as used here is the one in effect before tree mortality and ageing are applied.

15. New Spring Foliage Production

$$A = \frac{F_T^{**}}{3.8} = \text{photosynthetic assimilation rate} \qquad (A.46)$$

The new foliage levels are

$$F_1' = F_0 = A \qquad (A.47)$$

$$F_2' = (1 - .04 \cdot F_1^{**}) \cdot F_1^{**} + (3.17 - 2.51 \cdot A) \cdot A \cdot F_2^{**} \quad \text{(A.48)}$$

(Fig. 7)

Additionally, F_1' and F_2' are supplemented by the contribution of SA_{21} (which carries full foliage) entering the susceptible age class.

$$F_1(t+1) = \frac{SA_{21} \cdot (1.0) + SA \cdot (F_1')}{SA_{21} + SA} \quad \text{(A.49)}$$

$$F_2(t+1) = \frac{SA_{21} \cdot (2.8) + SA \cdot (F_2')}{SA_{21} + SA}$$

$$F_T = F_1 + F_2 \quad \text{(A.50)}$$

16. Calculation of σ_i, μ_i, and q_i

A. Surface area by age class i; σ_1

σ_i is linearly interpolated between the following values:

i	0	20	40	60	80	
σ_i	0	17.5	24	27	29.5	x 1000 tsf/acre

e.g., σ_1 = 875 tsf/acre

We also set σ_{75} = 29,500 tsf/acre. (Fig. 5)

B. Age specific susceptibility factor, μ_i

$\mu_i = 0$ $\quad i \leqq 21$

$\mu_i = 0.5$ $\quad 22 \leqq i \leqq 49$

$\mu_i = .02 \cdot (i-49) + 0.5$ $\quad 50 \leqq i \leqq 74$

$\mu_{75} = 1.0$

C. Timber volume by age; q_i

$$\text{Total volume} = Q = \sum_{i=22}^{75} q_i \cdot T_i \text{ (cubic feet/acre)}$$

q_i is linearly interpolated between

i	0	10	20	30	40	50	60	70	80
q_i	0	0	100	700	1400	2000	2350	2525	2700

$q_{75} = 2700$

GLOSSARY OF VARIABLE NAMES

(ARRANGED FUNCTIONALLY)

TEXT	FORTRAN	DEFINITION
		.... SUBROUTINES & FILES ....
	INSECT	BUDWORM SURVIVAL SUB-MODEL
	FOREST	FOREST RESPONSE SUB-MODEL
	CONTRL	CONTROL POLICY SUB-MODEL
	INDEX	OUTPUT INDICIES SUBROUTINE
	STATS	BIOLOGICAL INDICATORS SUBROUTINE
	START	INITIALIZATION SUBROUTINE
	SLP	LINEAR INTERPOLATION FUNCTION
	BW.DATA	DATA FILE FOR PARAMETER VALUES
	BW.WDAT	FILE OF WEATHER SEQUENCE OPTIONS
	BW.TDAT	FILE OF INITIAL TREE AGE SPECTRA
	INSPEC	FILE OF INCOMING EGG DISTRIBUTION
N_E	XNE	DENSITY OF EGGS/TSF OF BRANCH SURFACE AREA
	XNS	DENSITY OF SMALL LARVAE/TSF (INSTAR I & II)
	XNL	DENSITY OF LARGE LARVAE/TSF (INSTAR III - VI)
	XNP	DENSITY OF PUPAE/TSF
	XNA	DENSITY OF ADULTS/TSF
	XNF	DENSITY OF FEMALES/TSF
S_E	SE	SURVIVAL OF EGGS TO SMALL LARVAE (0.81)
	SS	SURVIVAL OF SMALL TO LARGE LARVAE
	SL	SURVIVAL OF LARGE LARVAE TO PUPAE
	SP	SURVIVAL OF PUPAE TO ADULTS
	SF	SURVIVAL OF FEMALES TO EGG LAYING (1.0)
d_0	DO	MAXIMUM FOLIAGE RATION PER LARGE LARVAE (0.0074)
d_1	D1	LARVAL CONSUMPTION OF NEW FOLIAGE (F1)
	DR1	RATION OF F1 RELATIVE TO DO
d_2	D2	LARVAL CONSUMPTION OF OLD FOLIAGE (F2)
	DR2	RATION OF F2 RELATIVE TO DO
	PREFF2	PREFERENCE FACTOR FOR F2 OVER F1 (1.0)
	DR	TOTAL FOLIAGE RATION RELATIVE TO DO

	DEF1	DEFOLIATION OF F1, NEW FOLIAGE REMOVED BY LARVAE
	DEF2	DEFOLIATION OF F2, OLD FOLIAGE REMOVED BY LARVAE
	DEF	TOTAL DEFOLIATION (IN FOLIAGE UNITS)
S'_L	SLPAR	SURVIVAL OF NL FROM PARASITISM
	PAR1	MAXIMUM PARASITISM LEVEL
	PAR2	PARASITISM "DILUTION COEFFICIENT"
k_L	SLO	BASE SURVIVAL OF NL
N_6	XN6	DENSITY OF INSTAR VI/TSF
	PREY	EFFECTIVE DENSITY OF N6
	PRMAX	MAXIMUM PREDATION MORTALITY
α_i	PRALF(I)	)
A_i	PRA1(I)	) PARAMETERS FOR PREDATOR FUNCTIONAL RESPONSE
B_i	PRA2(I)	)
	PPED	TOTAL INSTANTANEOUS FUNCTIONAL RESPONSE
	PRFAC	SCALING FACTOR MULTIPLYING PREDATION INTENSITY
	AK	PREDATOR COMPETITION COEFFICIENT
S_{PRED}	SPRED	SURVIVAL OF NL FROM PREDATION
A_P	SPA	SP FUNCTION INTERCEPT
B_P	SPB	SP FUNCTION SLOPE
P_F	PF	PROPORTION OF FEMALES
A_{PF}	PFA	PF FUNCTION INTERCEPT
B_{PF}	PFB	PF FUNCTION SLOPE
	WGT	WEIGHT OF PUPAE
A_{F1}	FA1	PUPAL WEIGHT ATTRIBUTED TO F1 CONSUMPTION
A_{F2}	FA2	PUPAL WEIGHT ATTRIBUTED TO F2 CONSUMPTION
B_F	FB	PUPAL WEIGHT INTERCEPT
	FECA	FECUNDITY FUNCTION SLOPE (VS. WGT**1/3)
	FECB	FECUNDITY FUNCTION INTERCEPT
	FECLMT	LOWER LIMIT OF FECUNDITY
e_g	EGGS	DENSITY OF EGGS/TSF GENERATED ON SITE
F_{MIG}	FMIG	FRACTION OF POTENTIAL EGGS THAT MIGRATE OFF SITE
	FMGK(1)	MINIMUM FRACTION MIGRATING OFF SITE
	FMGK(2)	MAXIMUM FRACTION MIGRATING OFF SITE
	FMGK(3)	FMIG FUNCTION INTERCEPT
	FMGK(4)	FMIG FUNCTION SLOPE
e_s	EGGSTA	DENSITY OF EGGS/TSF STAYING ON SITE
E_0	EGGOUT	MEASURE OF EGG OUT-MIGRATION/LAND ACRE
	DP	SUM (AVERAGE) OF OUT-MIGRANT LEVELS
	DDP	SUM OF SQUARES OF OUT-MIGRANT LEVELS
	DSD	STANDARD DEVIATION OF EGG OUT-MIGRATION SPECTRUM
	SPEC(.)	SPECTRUM (LOG) OF OUT-MIGRATION LEVELS
	ISPEC(.)	INTEGER VERSION OF "SPEC"

Symbol	Variable	Description
R_{in}	REIN	FRACTION OF EGGS LEAVING A SITE THAT RETURN
E^*	EGFLUX	EXTERNAL EGG SOURCE (CONSTANT)
	SPECIN(.)	SPECTRUM FOR PICKING STOCHASTIC INFLUX LEVELS
E_i	EGGIN	EGGS/ACRE ENTERING FROM OFF SITE
	DENFAC	DENSTIY FACTOR TO CONVERT "EGGIN" TO #/TSF
e_p	PEGGS	TOTAL DENSTIY OF EGGS 'SEEKING' OVIPOSITION SITES
P_1	P1	FOLIAGE SEARCH PROBABILITY FUNCTION
	EBASE	LOWER LEVEL OF EGG DENSITY (1.0XE-5)
P_2	P2	FOLIAGE SEARCH PROBABILITY FOR INSTAR II
P^{ss}	PSS	COMPOSITE SEARCH PROBABILITY FOR SMALL LARVAE
k_s	SSO	BASE SURVIVAL OF NS
	IW	WEATHER INDEX: 1, 2, 3 (BAD, MED, GOOD)
	IWTHR(.)	SYNTHETIC WEATHER TRACE (UP TO 1000 YEARS)
W_F	WFAC(IW)	FACTOR RELATING WEATHER TO LARGE LARVAL SURVIVAL
	IW1	IW(NYR-1)
	IW2	IW(NYR+1)
	WA3	3-YEAR MOVING AVERAGE OF WEATHER
T_i	T(I)	PROPORTION OF HOST LAND WITH TREES IN AGE GROUP I
	T1	SUM OF T(I) (ALWAYS = 1.0)
T_{sp}	TSP	SUSCEPTIBLE FRACTION OF LAND (TREES > 21 YEARS)
T_{21}	T21	FRACTION OF FOREST ENTERING SUSCEPTIBLE AGE
	MSAGE	MINIMUM AGE FOR SUSCEPTIBILITY
	ITRAN	TRANSITION AGE INTO SUSCEPTIBLE CLASS
	POLD	FRACTION OF HOST LAND WITH OLD TREES (> 50 YEARS)
	IOLD	LOWER CUT-OFF FOR OLD TREE CATEGORY
F_1	F1	AMOUNT OF NEW FOLIAGE/TSF (ARBITRARY "FOLIAGE UNITS")
F_2	F2	AMOUNT OF OLD FOLIAGE/TSF
F_T	FT	TOTAL FOLIAGE/TSF
	FLOSS	FRACTION OF MAXIMUM FOLIAGE MISSING
F_1^{**}	F1D	)
	F2D	) FOLIAGE LEVEL AFTER FEEDING ("AUTUMN")
	FTD	)
F_1^*	F1H	)
	F2H	) TEMPORARY FOLIAGE LEVELS BEFORE MORTALITY COMPENSATION
	FTH	)
	FDRAT	FOLIAGE RATIO ON "DYING" TO "LIVING" TREES
	FOLUP	FOLIAGE COMPENSATION FACTOR FOR TREE MORTALITY
$*_{sp}$	TSP1	"TSP" BEFORE FOREST UPDATE
$_L$	FRLV	FRACTION OF SUSCEPTIBLE LAND THAT SURVIVES MORTALITY

A	F0	PHOTOSYNTHETIC ASSIMILATION RATE
	FOK	COEFFICIENT OF PHOTOSYNTHETIC ASSIMILATION RATE
	F1K	BASIC LOSS RATE OF NEW FOLIAGE
	F2K	COEFFICIENT OF F2 SURVIVAL
	F2KK	COEFFICIENT OF F2 SURVIVAL
σ_i	SIG(I)	BRANCH SURFACE AREA OF AGE GROUP I (TSF/ACRE)
	SIGP(.)	INTERPOLATOR POINTS FOR SURFACE AREA CURVE
SA	SURF	BRANCH SURFACE AREA OF SUSCEPTIBLE TREES (> 21 YEARS)
	SAR	BRANCH SURFACE AREA INDEX (SURF/24000)
SA_{21}	SA21	AVERAGE SURFACE AREA OF AGE GROUPS 20, 21 AND 22
	TMNAT	"NATURAL" TREE MORTALITY RATE
RG_{75}	REGNAT	REGENERATION FROM "NATURAL" MORTALITY
S_{BW}	STRESS	STRESS FACTOR DUE TO ACCUMULATED DEFOLIATION
S_0	SBWO	MAXIMUM STRESS
μ_i	XMU(I)	BUDWORM DAMAGE SUSCEPTIBILITY OF AGE CLASS I
M_i	TM	TREE MORTALITY RATE FOR AGE CLASS I
RG_{BW}	REGBW	REGENERATION (TO CLASS 1) DUE TO BUDWORM ARBORCIDE
	CHOMP	FIBRE VOLUME DESTROYED BY BUDWORM
	REG	TOTAL REGENERATION FROM ALL CAUSES
R_{SA}	RSA	SURFACE AREA REDUCTION FACTOR DUE TO BUDWORM DAMAGE
	RSAK	MULTIPLIER OF STRESS TO GET RSA
	RSAMAX	MAXIMUM OF RSA
	ISWCH(I)	SWITCHES TO IMPLEMENT VARIOUS PROGRAMME OPTIONS
	NTOT	TOTAL NUMBER OF YEARS OF SIMULATION
	NYR	YEAR NUMBER OF SIMULATION
	NWTH	INDEX NUMBER FOR WEATHER SEQUENCE SET
	NTR	INDEX NUMBER FOR INITIAL TREE AGE SPECTRUM
	NFOL	INDEX NUMBER FOR INITIAL FOLIAGE CONDITION
	TO(I)	INITIAL TREE AGE SPECTRUM
	F10	)
	F20	) INITIAL FOLIAGE
	FTO	)
	XNLO	INITIAL LARGE LARVAL DENSITY
	"---NU"	VALUES OF VARIABLES AT THE END OF EACH ITERATION
P_H	PHOST	FRACTION OF LAND SUPPORTING HOST TREE SPECIES
	ACRES	NUMBER OF ACRES PER SITE (42100)
q_i	QU(I)	FIBRE VOLUME FOR AGE CLASS I (CUBIC FEET/ACRE)
	QUP(.)	INTERPOLATOR POINTS FOR VOLUME CURVE
Q	QVOL	TOTAL VOLUME/ACRE ON HOST LAND

TAGE	AVERAGE TREE AGE
TSD	STANDARD DEVIATION OF TREE AGE
HINF	TREE AGE DIVERSITY INDEX ("INFORMATION" INDEX)
IPOL	POLICY NUMBER
LOGAGE	AGE FOR LOGGING
CUT	FRACTION OF HOST LAND LOGGED
CLRCUT	ACRES LOGGED PER SITE (THOUSANDS)
POTLOG	PROPORTION OF TOTAL VOLUME LOGGED
HARVQ	VOLUME HARVESTED PER YEAR (MILLIONS)
PROFQU	PROFIT PER CUBIC FOOT (SANS CUTTING COSTS)
PROFIT	PROFIT PER YEAR (MILLIONS $)
SALEP	SALE PRICE PER CUBIC FOOT
STUMP	STUMPAGE CHARGES PER CUBIC FOOT
COSTTR	COST OF TRANSPORTATION PER CUBIC FOOT
COSTLG	COST OF LOGGING PER CUBIC FOOT
AGESCT(I)	COST OF LOGGING AGE CLASS I ($/C.F.)
CLOGP(.)	INTERPOLATOR POINTS FOR COST OF LOGGING CURVE
HAZ	HAZARD INDEX (NEW BRUNSWICK VERSION)
HAZTH	HAZARD INDEX THRESHOLD TO INITIATE SPRAY POLICY
EGTH6	EGG THRESHOLD FOR SPRAYING IN THREAT STATE
SATH6	SA THRESHOLD FOR SPRAYING IN THREAT STATE
HZEG(.)	EGG MASS LEVELS FOR HAZARD INDEX
IHZE	HAZARD INDEX FOR EGG DENSITY
HZF1(.)	F1 LEVELS FOR HAZARD INDEX
IHZF1	HAZARD INDEX FOR F1 LEVEL
HZF2(.)	F2 LEVELS FOR HAZARD INDEX
IHZF2	HAZARD INDEX FOR F2 LEVEL
HZFT(.)	FT LEVELS FOR HAZARD INDEX
IHZFT	HAZARD INDEX FOR FT LEVEL
DOSHAF(I)	SPRAY DOSAGE FOR 50% MORTALITY
DOSE(I)	DOSAGE SPRAYED (IN MODE I) IN ANY YEAR
COSTDS(I)	COST OF SPRAYING A 50% KILL DOSEAGE IN MODE I
COSTSP	COST OF SPRAYING A SITE
ISPRAY(I)	INDEX TO DESIGNATE OPITION TO USE SPRAY MODE I
IACCUM(I)	TOTAL NUMBER OF YEARS THAT MODE I WAS IMPLIMENTED
SPRAY(I)	TARGET MORTALITY LEVEL FOR SPRAY MODE I
IFLAG	YES OR NO FLAG FOR SPRAYING IN CURRENT YEAR
ZAP(I)	LEVEL OF MORTALITY APPLIED (MODE I)
FOLSAV	FRACTION OF FOLIAGE SAVED BY SPRAYING LARVAE

THE ENVIRONMENTAL MANAGEMENT UNIT DEPARTMENT OF ZOOLOGY AND APPLIED ENTOMOLOGY, IMPERIAL COLLEGE OF SCIENCE & TECHNOLOGY, SILWOOD PARK, ASCOT, U.K.

INTRODUCTION

The Unit was set up in 1970 under a grant from the Office of Resources & Environment in the Ford Foundation to develop analytical tools and management techniques for environmental resource problems. A major part of the research effort has been focussed on problems of pest management. The original members of the Unit had experience in this field; the parent department of Zoology and Applied Entomology at Imperial College has a well established research program in many aspects of pest control; and, as experience has shown, pest management problems provide an ideal framework for developing the tools and techniques which are applicable to a variety of other problems in the broader field of environmental management. Thus, in recent years, the Unit has also pursued research on a variety of other topics, including stability analysis of ecological systems, economics of pollution and resources, methodologies for ecosystem and river basin studies and the environmental problems of developing countries.

Our research strategy has been to focus on a selected number of case studies using our experiences to develop concepts and tools of general applicability. Table 1 lists the major case studies, either completed or currently underway. In each case the Unit is collaborating with a particular research organisation and with one or more individuals who are intimately involved with the research and management aspects of each case study. The Unit thus tends to operate at one stage removed from the actual problem, providing research methodology, ecological and economic analysis, a variety of computer simulation models and operations research techniques, as an aid to the scientists and managers with direct responsibility for the particular problem. The work pattern of the Unit relies heavily on seminars and workshops to identify overall objectives for each case study, to provide a continuing process of reassessing objectives, and to integrate different aspects of the work. Most of the detailed research, however, is carried out by members of the Unit working in twos and threes or by individuals working alone. The seminars thus provide a means of cohesion and a forum for exchange of ideas, but individuals are free to pursue new ideas providing that they fall under the general heading of the research program.

Table 1. E.M.U. Case Studies.

Project	Location	Organization	Co-workers
Aedes-yellow fever virus system	East Africa	WHO	G.A.H. McClelland
Sugar cane frog-hopper	Trinidad	Caroni Ltd.	A.B.S. King
Cattle tick	Queensland, Australia	C.S.I.R.O.	R.W. Sutherst
Leaf cutting ant	Trinidad	ODM	T. Lewis
Cotton bollworm	Malawi	ODM	J. Marks J. Farrington
Savanna ecosystem	South Africa	C.S.I.R. Witwatersrand	B. Walker
ADEMA river basins	South America	UNEP/ECLA	M. Nelson E. O'Connell

The case studies have not been chosen at random but selected as far as possible to provide a logical growth of our knowledge of the application of systems analysis techniques to environmental problems. In particular our pest case studies have been chosen with the aim of developing a coherent theory of pest management in the group. To this end, we have attempted to identify the major paradigms in pest control and chosen case studies that illustrate these. Early on in our work it became apparent that major advances in this field required a firm integration of both ecological and economic analysis. We have also found that this integration is best furthered by adopting a classical decision theory framework and concentrating initially on the simplest pest control paradigms for which there are corresponding real-world pest control problems. In this sense our research has been somewhat incremental in character, the growing theory developing on a basis of building blocks made from our analyses of increasingly more complicated paradigms and case studies.

The current staff of the Unit consists of the following: G.R. Conway - Reader in Environmental Management, with a Ph.D. in Systems Ecology, currently working on models of pest population dynamics and control and on research and training methods for environmental problems in developing countries.

G.A. Norton - Lecturer in Resource Economics, with a Ph.D. in Agricultural Economics and currently working on the analysis of decision making in crop protection and the economics of pollution and resource use.

H.N. Comins - Research Fellow with a Ph.D. in Theoretical Physics, supported by a Fellowship from the Natural Environmental Research Council, and working on theoretical studies of population competition, insecticide resistance and game theory applied to resource problems.

M. Birley - Research Officer with a M.Sc. in Control Theory, completing a Ph.D. in Ecology, and working on a transfer function approach to the problems of estimation and control of pest populations.

J. Parlour - Research Fellow, with a Ph.D. in Agricultural Economics, who formerly worked for the Ministry of Urban Affairs in Canada and is currently working on the methodology of systems analysis for social systems as they relate to the understanding and management of environmental problems.

N. Barlow - Research Assistant, completing a Ph.D. in Ecology, currently working on models of the dynamics of aphid populations.

S. Patton - Research Assistant, completing a Ph.D. in Zoology, currently working under a contract from the U.K. Department of the Environment to include the effects, both positive and negative, of pesticide use on food production in the Department's global resources model.

I. Craig - Research Assistant, with a B.Sc. in Agriculture, also working on the Department of Environment contract.

BACKGROUND TO AGRICULTURAL PEST MANAGEMENT MODELLING

G. A. Norton

INTRODUCTION

It is a number of years now since systems analysis and modelling techniques were first seen to have a potential role to play in pest management. Nevertheless, I think it is true to say that the inspirational source of most of the papers published on the subject can still be traced to one man - Ken Watt in California. Indeed, as Conway (1973) has pointed out, prior to 1970, Watt had made virtually the only contribution to applied systems analysis in pest management. And yet, in 1970, Watt (1970) himself confessed that there was "... almost no visible impact of pest modelling on current control practice". Despite the flurry of modelling activity that has taken place since this time, I think it fair to say that the situation has changed little since then. But in saying this, the question then arises, what type of contribution to pest management can we expect systems analysis and modelling to make?

In other fields of resource and environmental management, the rationale for systems analysis and modelling can be grounded on the concept of 'hypotheticality' (Haefele, 1974). The argument goes like this. As one cannot or does not wish to test a whole range of ecological management strategies in the real-world, then the impact of each potential strategy remains hypothetical. However, it is often essential that a choice be made, despite the absence of empirical investigation. It is here that systems analysis and modelling techniques can be of value, providing a rigorous means of hypothetically investigating the impact of each strategy and, we hope, increasing the chances of deciding on the best one.

In pest management this hypotheticality argument is much weaker. Indeed, as Way (1973) has pointed out, the development of real-life strategies in the past has been largely based on the results of hypothesis testing through empirical experiments in the field. Consequently, as might be expected, views concerning the potential value of systems analysis and modelling in a pest management context are diverse, ranging from blanket acceptance of the approach (Stark and Smith, 1971) to the view that systems analysis is only justified where empirical attempts at improving management have failed (Clark, 1970).

On the premise that pest management modelling is indeed concerned with the improvement of pest management, the aim is to show how this goal might be achieved. The paper is divided into two parts. The first part investigates the nature of the pest management problem, while in the second part, attention is focussed on the role that modelling has to play. In concluding, a more pragmatic application of systems analysis and modelling to pest management is proposed.

THE AGRICULTURAL PEST MANAGEMENT PROBLEM

As with other natural hazards, such as floods and droughts, the management problem associated with pest attack arises from the fact that pests reduce the yield and quality of crop produce. In analyzing pest problems, it is useful to distinguish between two types of pests according to the nature of attack. The first category, exogenous pests, enter the crop for only part of their life-cycle and bear most similarity to other natural hazards. Locusts and wind-borne diseases are good examples. The second category, endogenous pests, including nematodes and weeds, complete all, or virtually all, of their life-cycle within the crop and tend to build up over time.

Apart from the nature of pest attack and the damage it causes, the dimensions of a pest hazard also depend upon the extent to which cropping systems are adapted to pest attack or regulation measures can be taken in response to it. In subsistence agriculture, the emphasis is on adaptation, protection measures being an intrinsic part of the farming system. At its simplest, this adaptation involves planting a diversity of crops over space and time, spreading the risk of crop loss. With further investigation of these cropping systems, however, it appears likely that adaptation is more subtle than this; various features of the cropping system being managed to reduce the attractiveness of the crop or its favorability to insect attack (Norton, 1975).

In commercial agriculture, where monocultures are the rule, farmers employ technological rather than cultural methods of crop protection (Ogunfowora and Norman, 1973). In these circumstances, the farmer's decision tree (Fig. 1) invariably involves pesticides, although such alternatives as biological control agents or juvenile hormones may also be included. This state of affairs, where heavy reliance is placed on pesticides, is well described by Cyert and March (1963) who, in fact, were describing the behavior of industrial firms, when they say that "... firms achieve a reasonably manageable decision situation by avoiding planning where plans depend on predictions of uncertain future events and by emphasizing planning where the plans can be made self-confirming through some control device".

Decision Analysis

The purpose of this overview has been to illustrate that pest management decision making is determined by three factors:

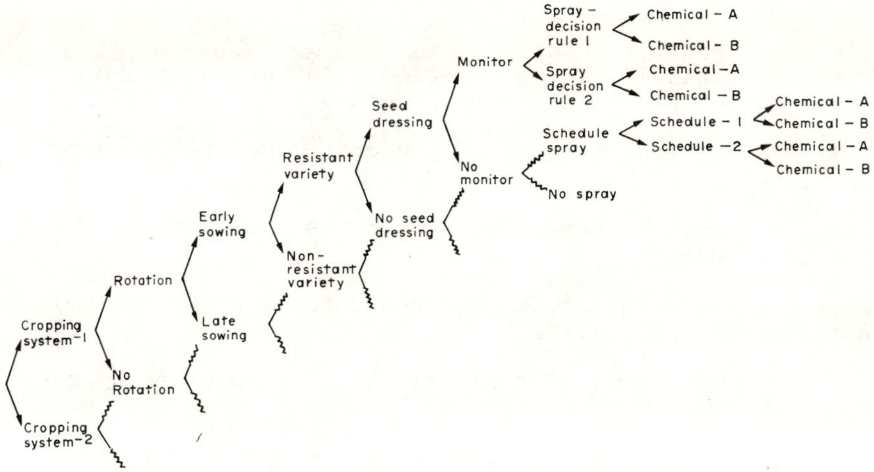

Figure 1. Hypothetical pest management decision tree.

1. the nature of pest attack and the damage it causes;

2. the range of protection measures and information available to the farmer; and

3. the farmer's objectives.

But this does not indicate how a decision is arrived at. From a theoretical standpoint, the pest management decision problem can be tackled by means of a two-stage analysis: determining the best strategy to adopt for any dimension of attack, and estimating the dimension of attack faced at a particular time and place. To illustrate, let us consider a particular case study involving the control of potato cyst eelworm by means of a soil fumigant, DD.

In this particular example, the level of pest attack can be estimated by taking soil samples from the field to be planted, and determining the number of nematode eggs per gram of soil. On the basis of this information, the farmer can then decide whether to apply a chemical fumigant (DD) before planting his crop or not. In making an analysis, however, additional information is required.

(i) The damage function—relating the level of attack to crop loss. Brown (1969) has found that the damage function for eelworm is approximately linear, a loss of 0.1 tons/hectare (ha.) being associated with 1 egg/gram (gm.) of soil.

(ii) The control function—that relates the reduction in attack to the control strategy applied. In this case, DD achieves a reduction in attack of around 80% (Jones, 1973).

(iii) The estimated price of the crop. For the time being assumed at £40/ton.

(iv) The cost of the chemical and its application. This is taken as £100/ha.

With this information, we can now determine the economic threshold for DD control.

Where θ = the level of pest attack in terms of eggs/gm of soil,

d = the damage coefficient in tons/ha. lost for each egg/gm of soil,

k = the mortality coefficient associated with DD,

p = the price of potatoes,

c = the cost of applying DD,

then the loss in revenue associated with eelworm attack

$$= pd\theta$$

and the reduction in loss associated with applying DD

$$= pd\theta k \quad .$$

Hence, it will be profitable to apply DD where

$$pd\theta k \geq c$$

that is, the economic threshold (θ^*) is where

$$\theta^* = \frac{c}{pdk} \quad .$$

From the information given above,

$$\theta^* = \frac{100}{40 \times 0.1 \times 0.8} = 31 \text{ eggs/gm. of soil.}$$

Clearly, this threshold will change as its components change. For instance, we can consider the effect of changes in the price of potatoes and the cost of control, or changes in the damage and control function, associated with the use of different varieties or new nematicides. It will also be evident that this type of decision rule, based on an economic threshold model, can only be used where an estimate of pest attack is available. In the case of the potato cyst eelworm, the monitoring of eggs in the soil provides a valid measure of pest attack for the following reasons:

(a) the eelworm is an endogenous pest, with negligible migration into the crop,

(b) the climate of the soil in Britain is relatively uniform,

(c) the eelworm attacks a single host crop, potatoes, and

(d) it has no significant predators or parasites.

In most cases, pest attack will be more complex than this. Some exogenous pests, such as locusts, can attack a crop with little or no warning, often travelling considerable distances to do so. Moreover, the development of attack may be affected by such factors as climate or natural enemies. Hence, although it may be possible to determine the best strategy to adopt for a given level of attack, the second stage of analysis, to determine the particular level of attack, cannot be completed. In this case, prophylactic measures have to be applied, the control decision being made without an estimate of pest attack.

This does not mean, however, that the control decision has to be made in an arbitrary fashion. Since the farmer will have some idea of the probability distribution of attack, an estimate can be made of the expected outcome and its probability distribution for each strategy. The case of potato blight and its control by the application of schedule sprays serves to illustrate this type of analysis.

Following Large (1958), four levels of attack are identified, according to the time at which 75% of the haulm is blighted. On the basis of historical records, Table 1 shows the probability of these four levels of attack in four regions of the U.K. From various field trials, a generalized pay-off matrix can be constructed in terms of the yield of potatoes associated with each strategy (no spray/spray) at each level of attack (Table 2). A monetary pay-off matrix is easily derived from this (Table 3), and the expected monetary value of each strategy in each region can be found by weighting the outcome associated with each level of attack by the probability of this level of attack occurring (Table 4).

Decision Criteria

If farmers do opt for a particular strategy of blight control on the basis of its expected monetary value, then Table 4 indicates that it is best to spray in the South-west and the Fens, but not in the South or the North. But is expected value an appropriate criterion for making such decisions? This will only be the case where farmers are risk neutral and indifferent to the range of outcomes that occur. Most farmers are more likely to be risk averse, valuing initial increments in yield or revenue far higher than later increments.

Table 1. Probability distribution of blight attack (based on Large, 1958).

Region	Level of attack			
	θ_1 (Mid-Aug)	θ_2 (End of Aug)	θ_3 (Mid-Sept)	θ_4 (End of Sept)
South-west	0.5	0.5	0	0
Fens	0.4	0.2	0.1	0.3
Southern	0	0.5	0.2	0.3
Northern	0	0	0.5	0.5

Table 2. Physical pay-off matrix (tonnes/ha).

Level of attack	Strategy	
	Unsprayed	Sprayed
θ_1	22	26
θ_2	26	28
θ_3	29	29*
θ_4	30	29*

*3% loss by wheel damage.

Table 3. Monetary pay-off matrix (£/ha).

Level of attack	Strategy	
	Unsprayed	Sprayed
θ_1	880	1,010
θ_2	1,040	1,090
θ_3	1,160	1,130
θ_4	1,200	1,130

Table 4. Expected monetary values (£/ha).

	Strategy	
	Unsprayed	Sprayed
South-west	960	1,050*
Fens	1,036	1,074*
Southern	1,112*	1,110
Northern	1,180*	1,130

*best strategy

Consider a risk averse grower in the Southern region who incurs other costs of production, amounting to £1,060/ha. From Tables 1 and 3, we can determine the probability distribution of outcomes, as shown in Table 5. Although the expected value of the no-spray strategy is higher, it is also associated with a 50% chance of making a loss. Consequently, the farmer will not hesitate to adopt the less profitable, but less risky, option of spraying. Subsistence farmers are likely to be even more risk adverse.

Table 5. Probability distribution of crop income in the Southern region.

Unsprayed		Sprayed	
Outcome (£s/ha.)	Probability	Outcome (£s/ha.)	Probability
-20	0.5	-	-
-	-	+30	0.5
-	-	+70	0.5
+100	0.2	-	-
+140	0.3	-	-

Risk not only arises where there is a lack of information on the current level of attack, however. Even where pest attack is monitored—as in the eelworm example—risk or, more properly, uncertainty can arise from sampling errors, from inadequate information on the damage and control function or the effect of climatic and agronomic factors on these functions, as well as from poorly estimated prices and costs. Consequently, decision making in agricultural pest management will often take place where the farmer has only the vaguest idea of the probability distribution of outcomes associated with each strategy available to him. In these circumstances, a choice of control strategy will be arrived at through a process of trial and error, farmers adopting those strategies that have proved satisfactory in the past.

In some cases, this satisficing approach to decision making may obtain the same result as an optimal solution. This can occur where there are severe risk constraints, where the protective measure involves on all-or-nothing decision, where the optimal solution is insensitive to factor changes, due to relatively low pesticide costs, for example, or where the cost of obtaining information for optimization is prohibitive. In many other cases, however, there is likely to be considerable room for improving pest management. It is here where modelling may be of help.

MODELLING AND AGRICULTURAL PEST MANAGEMENT

Having reviewed the essential ingredients of agricultural pest management problems in a decision making context, let us continue in this vein and look at the modelling problem in the same terms. As with pest management itself, the decisions made in constructing pest management models will be determined by three main factors: the nature of the pest management problem, the models and information available, and the objectives of the modellers. Let us consider each of these in turn.

I. Nature of the Pest Management Problem

In our earlier examination of the potato cyst eelworm, we found that a solution to the decision problem—whether to apply DD or not—could be readily achieved if information on the number of eggs/gm of soil were available. The simplicity of this analysis can be attributed to five factors:

Pest attack and damage -

(a) Population development is predictable,
(b) The damage function is linear,

Control measures -

(c) A single control measure is considered, involving a binary (yes/no) choice,
(d) The control function consists of a single, mortality coefficient, and

Objectives -

(e) There is a single, objective criterion of profitability.

Using the nematode case as our point of departure, we can now consider how the three components of the pest management problem can increase in complexity, and where modelling techniques can be employed.

Nature of pest attack and damage

We have already seen that whereas the level of damaging attack of eelworm is a linear function of egg numbers, in most other pests the nature of attack is more complex. Not only may individuals enter and leave the crop (exogenous pests), but the development of pest attack may be influenced by the effect of climate on hatching and development, and by the operation of population regulation processes, including predation and parasitism. Depending on the pest considered, these complicating factors can be explicitly accounted for in population models by the use of degree-days (Shoemaker, 1977), density-dependence relationships (Conway, 1977), and predator-prey models (Kiritani, 1977).

The description of pest attack may be further complicated where its spatial and temporal aspects are important. Where mobile (exogenous) pests are considered, particularly in connection with planation crops or regional control schemes, the assumption of spatial homogeneity of attack, implicitly made for the nemotode example, will be unacceptable. Similarly, where less mobile (endogenous) pests build up from season to season, the temporal dimensions considered need to be extended, as we shall see later in reconsidering the nematode control decision.

For many pest situations, the assumption of a linear damage function will also be inappropriate where significant changes in the damage coefficient occur with variations in the level and time of attack. Again, attempts have been made to account for this complexity in modelling programs (Gutierrez and Wang, 1977) although, in general, I would suggest that this is one area where empirical investigation takes priority, at least for the present.

The control measures considered and their effectiveness

It was implicitly assumed in the case of potato cyst nematode that the decision whether to apply DD or not was for a discrete crop season and that we were only interested in the outcome for that season. For exogenous pests, such as potato blight, where no significant physical relationship exists between attack in one year and the next, this assumption might be quite realistic. However, as we have already seen, endogenous pests, such as nematodes, often build up over time, and farmers apply control measures not only to reduce damage in the current year but also to reduce attack in the subsequent year. The effect of considering a two-year period on the economic threshold for eelworm attack is shown in Fig. 2 (based on data obtained from Jones (1973)).

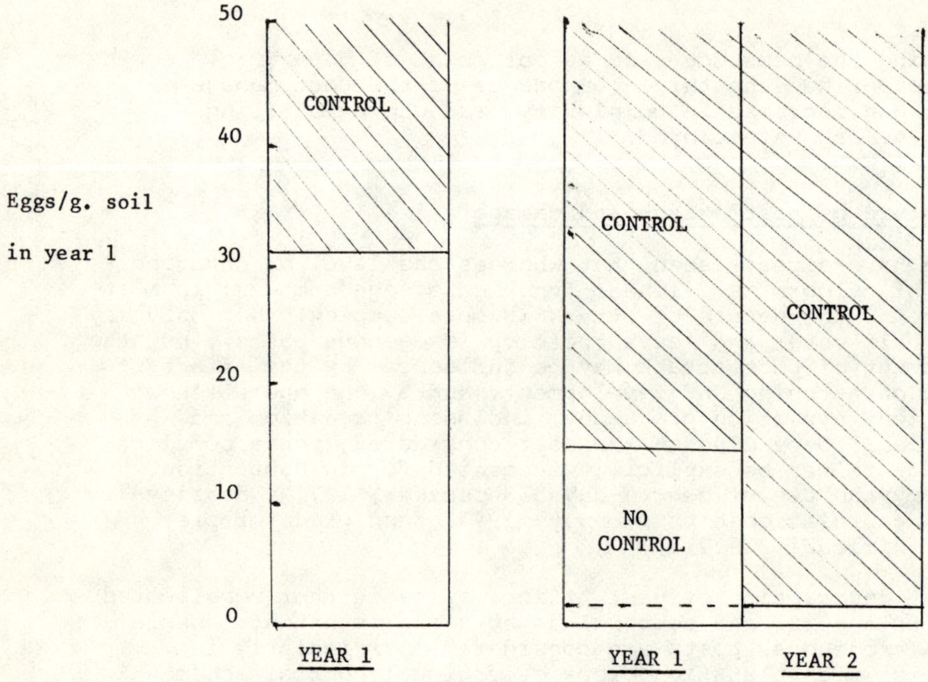

Figure 2. Single and two-year economic thresholder for potato cyst eelworm.

In practice, the potato grower's decision problem is likely to be more complex than this. Not only is his planning horizon likely to be greater than two years, but he will also be considering integrated control strategies, involving crop rotations and resistant varieties. Consequently, even in this simplest of agricultural pests we find that modelling techniques could prove useful in comprehensively "testing" long-term strategies of control.

But even with this more realistic view, the nematode case is still very simple since the control decisions remain binary (yes/no) options for each control measure. In other cases, where the number and time of control applications is variable, apart from strategic questions, such as which measures to employ, tactical questions concerning the degree and time of control also have to be answered. It is here where a single mortality coefficient, as used in the nematode example, is clearly unsatisfactory. In such cases, consideration has to be given to the determination of the control relationship in terms of the effect of the number and time of control applications on the level of pest attack. An example of the use of modelling in this role is given by Conway (1977).

Apart from their immediate effect on pest attack, control strategies may have a wider impact, especially where chemical methods of control are employed. Indirect impact on pest attack can occur through the effect of insecticides on natural enemies, whereas longer term impact on the effectiveness of control can occur through the development of insecticide resistance. Modelling has a role to play in both cases, particularly with "estimating" the outcomes of long-term strategies that affect pesticide resistance (Comins, 1977).

Objectives of the farmer

Aside from using profitability as our criterion for deciding whether to apply DD or not in a single year, we have now seen that in some situations account must be taken of the variance or trends that occur from season to season. In practice, pest management decisions will be affected by other criteria as well. Nevertheless, in most cases the primary goals of farmers, whether subsistence or commercial, are likely to be relatively simple, based on three factors:

(i) the farmer's evaluation of outcomes, in terms of food or net revenue;

(ii) his attitude to risk and uncertainty; and

(iii) his time horizon and time preferences.

On the other hand, although these same criteria may be used by public agencies to assess pest management programs, in these circumstances account will also be taken of wider, social objectives, making the analysis far more complex.

II. Range of Models and Information Available

As well as being determined by the nature of the pest management problem itself, the form of modelling effort will also be constrained by the range of modelling techniques available. The choice of statistical, simulation, analytical or optimization models will depend on the current state of the art, but, more importantly, on the finance and facilities available and on the personal predilections and ability of the modeller. The availability of information relevant to the pest management problem might also be thought of as a constraint, although its importance will undoubtedly depend upon the purpose of the modelling exercise.

III. The Objectives of Modelling

At the outset of this paper, we took the improvement of pest management to be the prime objective of modelling programs. More explicitly, modelling can attempt to achieve this objective by performing two roles. The first is concerned with improving the perception of the problem—particularly by research and extension workers, by combining all its relevant facets within a unifying framework. At a case study level, modelling in this descriptive role serves as a means of coordinating research effort, illustrating how biological, ecological and decision variables interact, identifying the most 'sensitive' variables, and indicating where gaps in information exist. At a general level, this descriptive role of modelling provides an effective means of expounding general principles and theories of pest management.

The second role of modelling is concerned with more direct and more tangible means of improving pest management through research and extension programs. Since these programs attempt to improve pest management in three ways—by widening the range of control measures available, by improving information on pest attack, and by improving decision rules—we can investigate the normative role that modelling can play with respect to each.

Widen the range of measures available

The development of new methods of pest management, including new products and formulations, improved methods of application, biological control and regional schemes, increase the range of measures that can be used. If these measures are to be adopted by farmers, however, not only must they meet farmers' operational constraints but their performance must be an improvement on measures currently in use. Before allocating research funds to the practical development of novel methods of control, therefore, modelling techniques can be used to assess their potential and determine whether they might feasibly meet farmers' objectives. It is in this context that models have been used to 'test' sterile male and sex attractant schemes (e.g. Murdie and Campion, 1972).

Improve information on pest attack

With existing methods of control, an improvement in pest management can often be achieved by converting a prophylactic measure into an economic threshold measure. Apart from determining appropriate decision rules, which we shall discuss next, this requires improved information on pest attack, either in the form of a forecast or as monitored information. As in the previous section, modelling may have a role to play in assessing the value of this information, although this is probably best carried out by decision theory techniques (Norton, 1976).

A more useful role for models exists in generating the information itself. As a proxy for monitoring pest attack, simulation models can be of value in converting monitored climatic data into a useful estimate of pest attack (Shoemaker, 1977). On the other hand, where monitoring of pest attack occurs at a number of locations, modelling techniques can be used to coordinate and interpret this information at a regional scale. In forecasting pest attack, considerable reliance has been placed in the past on the use of statistical (regression) models to predict the level of attack. In the future, perhaps we can expect simulation techniques to play a more active role in this context.

Improve decision rules

Decision making in pest management is concerned with finding answers to two related questions:

(a) What method or methods of control should I adopt?

(b) At what level and at what times should I apply these measures?

In situations where dynamic factors are not important features of the pest management problem, simple deterministic or probabilistic decision models will most likely suffice. However, where dynamic relationships, operating within or between crop seasons, are relevant to the pest management problem, more complex models are required. We have already seen that this is the case where long-term economic threshold criteria or integrated control programs are to be devised for potato cyst eelworm. Other instances where modelling, and dynamic programming in particular, is used to determine optimal control strategies are given by Conway (1977), Shoemaker (1977) and Comins (1977).

Deriving optimal pest management strategies is one thing. Developing decision rules that can be used by farmers in real-life pest management, I would suggest, is another. There may be exceptions, such as large plantation crops or homogeneous crop areas, where modelling programs can be developed that give 'on-line' instructions to decision makers. In most cases, however, I think it more likely that the transfer of model output to farmers will take place through extension agents and advisory entomologists, utilizing existing decision rules, such as the economic threshold (Stern *et al.*, 1959). In this way, the prescription of the model can be modified and simplified to suit the particular conditions and requirements of individual farms or groups of farmers.

By whatever means model output is used, however, three requirements must be met (after Dillon, 1971)

1. The model must be acceptable to the decision maker (whether the farmer or an intermediary) as a valid representation of the real-world.

2. It must be able to cope with farmers' or extension agent's views concerning uncertain events--typically associated with climate, prices and technology.

3. Output of the model must be in a form that is amenable to managerial assessment, whether at a farm or regional level.

CONCLUSION

On the premise that modelling in pest management is concerned with improving pest management, this paper has attempted to show where such pragmatic impact might occur. However, to demonstrate that modelling can lead to better pest management does not, of course, mean that it will necessarily happen. All one can do is attempt to increase the likelihood of it happening. It is in discussing just this problem that I wish to conclude.

If truly representative models of complete pest "life-systems" or agro-ecosystems are constructed, then a more comprehensive approach to pest management should result. However, apart from the expense, in terms of time and money involved in collecting the necessary data (Way, 1973) and constructing the model (G.M. van Dyne, 1972), it might still be found that the model, as with some of the International Biological Program models, is unable to answer specific management questions.

In contrast to the comprehensive model approach, therefore, I would suggest that in agricultural pest management more emphasis needs to be placed on the use of systems analysis, with modelling _per se_ a useful, but not necessarily an essential, component. Drawing on our own experience and from a recent paper produced in the Unit (Walker _et al._, 1977), I would like to conclude by describing how this procedure might operate.

The first stage of the analysis involves a description of the problem from a decision analysis viewpoint. This entails looking at the farming system as a whole and specifically in terms of the factors affecting pest management decisions. Modelling may occur at this phase but only in the form of overall descriptive (flow diagram) models or simple decision or analytical models. From this descriptive phase, key management questions arise which in turn lead to ecological and biological questions. It is also at this stage where research and extension questions are raised.

In attempting to answer these explicit questions, simple, empirical experiments should be used whenever possible: for instance, in determining the effect of simple strategies or in determining damage functions. The value of modelling will be in answering those questions where a degree of 'hypotheticality' exists: for instance, in testing strategies that are long-term, complex, beset by considerable variability, or which involve, as yet, hypothetical methods of control.

It is by directing attention towards key management problems and making explicit hypotheses about their solution that this procedure may increase the chances of real-life improvements being achieved.

ACKNOWLEDGMENTS

I am grateful to colleagues in the Environmental Management Unit and the Department of Zoology and Applied Entomology for stimulating many of the ideas contained in this paper.

REFERENCES

Brown, E.B. (1969), Assessment of damage caused to potatoes by potato cyst eelworm, *Heterodera Rostochiensis Woll. Ann Appl. Biol.*, 63, 493-502.

Clark, L.R. (1970), Analysis of pest situations through the life systems approach, in *Concepts of Pest Management*, R.L. Rabb and F.E. Guthrie, eds., North Carolina State University, Raleigh, 45-58.

Comins, H.N. (1977), The control of adaptable pests (this volume).

Conway, G.R. (1973), Experience in insect pest modelling: a review of models, uses and future directions, in *Insects: Studies in Population Management*, P.W. Geier, L.R. Clark, D.J. Anderson, and H.A. Nix, eds., Ecological Society of Australia (Memoirs 1), Canberra, 103-130.

Conway, G.R. (1977), Case studies of pest control (this volume).

Cyert, R.M., and J.G. March (1963), *A Behavioral Theory of the Firm*, Prentice-Hall, Englewood Cliffs, N.J.

Dillon, J.L. (1971), Interpreting systems simulation output for managerial decision-making, in *Systems Analysis in Agricultural Management*, J.B. Dent and J.R. Anderson, eds., John Wiley, Sydney, 85-120.

Gutierrez, A.P. and Y. Wang (1977), Applied population ecology: models for crop production and pest management (this volume).

Haefele, W. (1974), Hypotheticality and the new challenges: the pathfinder role of nuclear energy, *Minerva*, 10, 303-323.

Jones, F.G.W. (1973), Management of nematode populations in Great Britain, *Proc. Tall Timbers Conf. Anim. Control Habitat Manag.*, 4, 81-107.

Kiritani, K. (1977), Systems approach to pest management of the green rice leafhopper (this volume).

Large, E.C. (1958), Losses caused by potato blight in England and Wales, *Pl. Path*, 7, 39-48.

Murdie, G., and D.G. Campion (1972), Computer simulation of red bollworm populations in control programs using sterile males and sex attractants, *Cotton Growing Rev.*, 49, 276-284.

Norton, G.A. (1975), Multiple cropping and pest control: an economic perspective *Meded. Fac. Landbouww. Rijks. Univ. Gent.*, 40, 219-228.

Norton, G.A. (1976), Analysis of decision making in crop protection, *Agro-Ecosystems*, 3, 27-44.

Ogunfowora, O. and Norman, D.W. (1973), An optimization model for evaluating the stability of mixed cropping systems under changing resource and technology levels, *Bull. Rural Econ. Sociol.* (Ibadan, Nigeria), 8, 77-96.

Shoemaker, C. (1977), Optimal management of an alfalfa ecosystem (this volume).

Stark, R.W. and R.F. Smith (1971), Systems analysis and pest management, in *Biological Control*, C.B. Huffaker, ed., Plenum Press, New York, 331-345.

Stern, V.M., R.F. Smith, R. van den Bosch, and K.S. Hagen (1959), The integrated control concept, *Hilgardia*, 29, 81-101.

van Dyne, G.M. (1972), Organization and management of an integrated ecological research program - with special emphasis on systems analysis, universities, and scientific cooperation, in *Mathematical Models in Ecology*, J.R. Jeffers, ed., 111-172.

Walker, B.H., G.A. Norton, G.R. Conway, H.N. Comins, and M. Birley (1977), A procedure for multidisciplinary ecosystem research with reference to the South African Savanna Ecosystem Project, *J. Appl. Ecol.* (in press).

Watt, K.E.F. (1970), The systems point of view in pest management, in *Concepts of Pest Management*, R.L. Rabb, and F.E. Guthrie, eds., North Carolina State University, Raleigh, 71-79.

Way, M.J. (1973), Objectives, methods and scope of integrated control, in *Insects: Studies in Population Management*, P.W. Geier, L.R. Clark, D.J. Anderson, and H.A. Nix, eds., Ecological Society of Australia (Memoirs 1), Canberra, 137-152.

CASE STUDIES OF PEST CONTROL

G. R. Conway

INTRODUCTION

In this paper I wish to summarize our work on two pest control case studies. One concerns the sugar cane froghopper (Aeneolamia varia saccharina), a pest of sugar cane in Trinidad and the other concerns the cattle tick (Boophilus microplus), a pest on cattle in Australia. Work on the froghopper has been carried out in collaboration with A.B.S. King of the Caroni Research Station in Trinidad and has been reported in Conway et al. (1975). Work on the cattle tick is being carried out in collaboration with R.W. Sutherst of the C.S.I.R.O. in Australia and at present is only partially complete.

Both case studies are concerned with controlling a seasonal, multi-voltine pest population, where heavy reliance is placed on insecticides as the main technique of control. In both instances, we have modelled population change using the Leslie Matrix model (Leslie, 1945, Usher, 1972). Here population change is given by

$$\underset{\sim}{n}_{t+1} = \underset{\sim}{M} \; \underset{\sim}{n}_t \qquad (1)$$

where $\underset{\sim}{n}_{t+1}$ and $\underset{\sim}{n}_t$ are the age class vectors of the population at time t+1 and t and the matrix

$$\underset{\sim}{M} = \begin{matrix} f(0) & f(1) & \cdots & f(k-1) & f(k) \\ p(0) & 0 & \cdots & 0 & 0 \\ 0 & p(1) & \cdots & 0 & 0 \\ 0 & 0 & \cdots & 0 & 0 \\ 0 & 0 & \cdots & 0 & 0 \\ 0 & 0 & \cdots & p(k-1) & 0 \end{matrix}$$

where f(x) and p(x) are the age specific fecundities and survivals.

The Leslie Matrix has the merit of describing population change in a neat and conceptually clear manner. Construction of the matrix depends upon three well defined sets of biological information:

Development time which determines the size of the matrix,
Fecundity which gives the values for f(x), and
Survival and migration, which give p(x).

THE SUGAR CANE FROGHOPPER (AENEOLAMIA VARIA SACCHARINA)

The froghopper is a small bug with sucking mouth parts which feeds on the sap of the sugar cane plant. Froghopper nymphs live in cracks in the soil and feed on the roots. However, the important damage is caused by the adults which feed on the uppermost leaves. Heavy attack results in 'blight', a necrotic condition of the leaves which greatly reduces photosynthetic capacity.

There are four generations or broods of froghoppers a year on cane, the first arising after the onset of the wet season which occurs in May or June. Later broods lay diapause eggs which survive the dry season from December to May and give rise to first brood nymphs of the succeeding season (Figure 1).

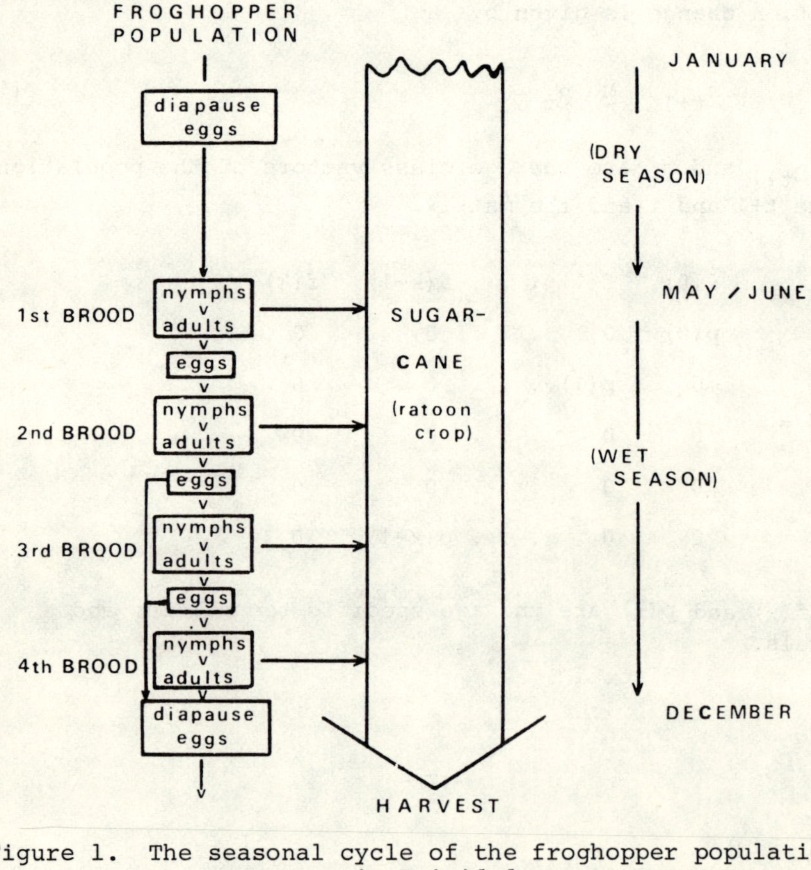

Figure 1. The seasonal cycle of the froghopper population on sugar cane in Trinidad.

Caroni Limited grows approximately 20,000 hectares of cane in Trinidad and current control practice is to spray infected fields from the air with an insecticide that kills the adult froghoppers. The fields are regularly monitored to determine the level of froghopper population, and spraying is carried out when the density exceeds 5 adults per 100 cane stems in the first brood, or 50 adults per 100 cane stems in later broods. The initial aim of our case study was to seek improvements in the economic efficiency of the current control practice of aerial spraying. We have assumed in this instance that the aim of control is to maximize profit. Hence, after Southwood and Norton (1973), the objective function can be described as

$$\underline{\text{Max.}} \quad Y[A(S)]\ P\ [A(S)] - C(S) \quad . \tag{2}$$

This expression, which relates cost of control to revenue, has within it four important functions:

(i) the cost of control function, C(S), which relates the cost of control to the control strategy S;

(ii) the control function, A(S), which relates the level of pest attack to the control strategy S;

(iii) the quantity damage function, Y[A(S)], which relates the yield to the level of attack A(S);

(iv) the quality damage function, P[A(S)], which relates the price to the level of attack A(S).

Our initial assumption, based on limited evidence, is that loss of sugar yield is a linear function of the froghopper population, expressed in terms of adult-days: that is, the yield of sugar per acre (Y) is given by:

$$Y = Y_o - dA(S) \tag{3}$$

where Y_o = the yield of sugar without froghopper attack = 3.5 tons per acre, and

d = the damage coefficient = 1.307×10^{-5} tons/acre/adult-day/200 canes.

More recent work suggests that in fact the relationship may be non-linear and that there may also be a significant effect of time of attack on sugar yield. However, because of the lack of detailed information we have continued the analysis on the basis of a simple linear damage function. The objective of improved control efficiency can thus be redefined as

$$\underline{\text{Max}}.\ P(Y_o - d\,A(S)) - C(S)\ . \qquad (4)$$

We have looked at the control function A(S) in two ways. First, we have attempted to understand the ecological characteristics of the pest which are relevant to control, using a population model based on the Leslie Matrix. Second, we have examined the dynamics of insecticide control, using a different population model and the technique of dynamic programming.

We began by taking a set of detailed field records for populations which had been unsprayed or only lightly sprayed such as that depicted in Figure 2. We then attempted to model the population pattern using a simple Leslie Matrix model with information obtained from the literature or from laboratory or field experiments. The model is assumed to have an age interval of one day and we further assumed that the values of p(x) and f(x) changed only with each new brood, a brood lasting from new laid eggs to the death of the adults to which they give rise. Each brood is thus described by a characteristic M matrix.

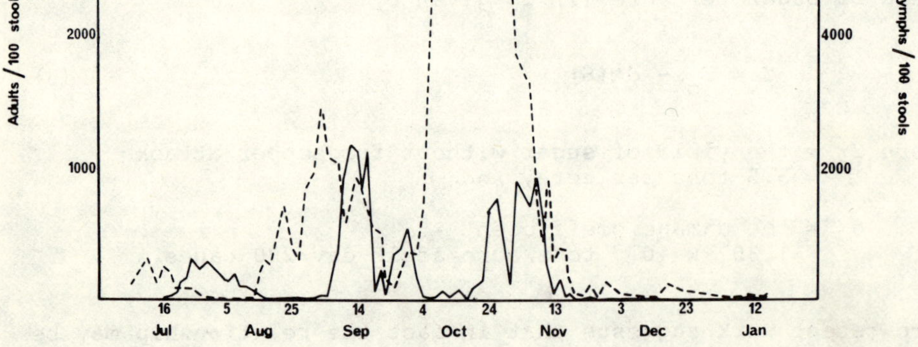

Figure 2. Record of the numbers of adult and nymphal froghoppers during the 1973-74 season for a single field (Orange Field 3) at Caroni Ltd. in Trinidad.

Because of the paucity of existing data at the beginning of the study, values for the parameters were obtained from a series of laboratory and field cage observations carried out during 1973 (Table 1). The estimates of fecundity and development could reasonably be assumed to approximate the values occurring naturally in the field; but the survival estimates are more likely to be over estimates. In fitting the model, we thus began by including the fecundity and development data and then, adjusting survival rates to provide a fit, comparing them subsequently with the available estimates.

Table 1. Estimates of development, fecundity and survival for the sugar cane froghopper obtained from laboratory and field cage experiments.

	Development times (days)				Fecundity		Survival (daily)		
Brood	Non-diapause[a] eggs (mean)	Nymphs (mode[b]) All	Male	Female	Eggs per Female	Proportion non-diapause eggs laid	Non-diapause eggs	Nymphs	Adults Female
1	19	28	28	32	13-38	0-932	-	?	0-894
2	20	28	28	32	12-93	0-703	0-981	09-35	0-838
3	24	28	28	32	10-77	0-253	0-983	?	0-803
4	27	28	28	32	13-26	0-321	0-990	?	?

[a]Non-diapause eggs are defined as eggs hatching in less than 35 days.

[b]Nymphal development times observed for the first brood are assumed to be similar for later broods

The model was initiated using a field estimate of the number of viable eggs present in May but it was assumed that all these eggs hatched on the 19th of June, when heavy rains first fell. Figure 3 shows the best fit to the female adult population. It was clearly an inadequate explanation of the brood pattern and we thus progressively added further components into the model. A reasonably realistic outcome was obtained with five additional components using data obtained from the laboratory, field cage or field experiments:

1. Hatching as a function of daily rainfall;
2. Fecundity increasing and then decreasing with female age;
3. Inclusion of the male froghopper population in the simulation, allowing for differing development rates of the sexes;
4. Simulation of the effect of insecticide spraying;
5. A reduction in egg development time, following re-examination of the data.

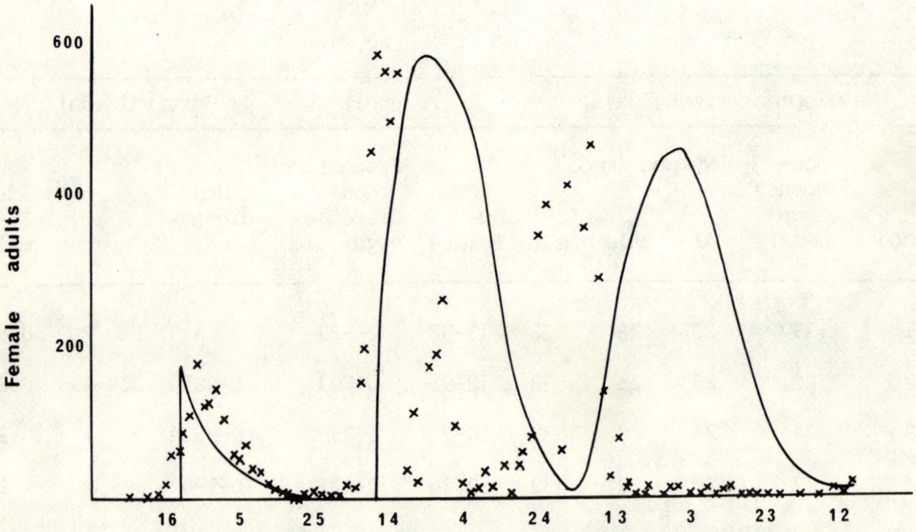

Figure 3. Preliminary fit of the model to female adult froghoppers recorded in Orange Field 3 in 1973-74

Figures 4 and 5 show the resulting fits to the adult and fifth instar nymph data for one particular field.

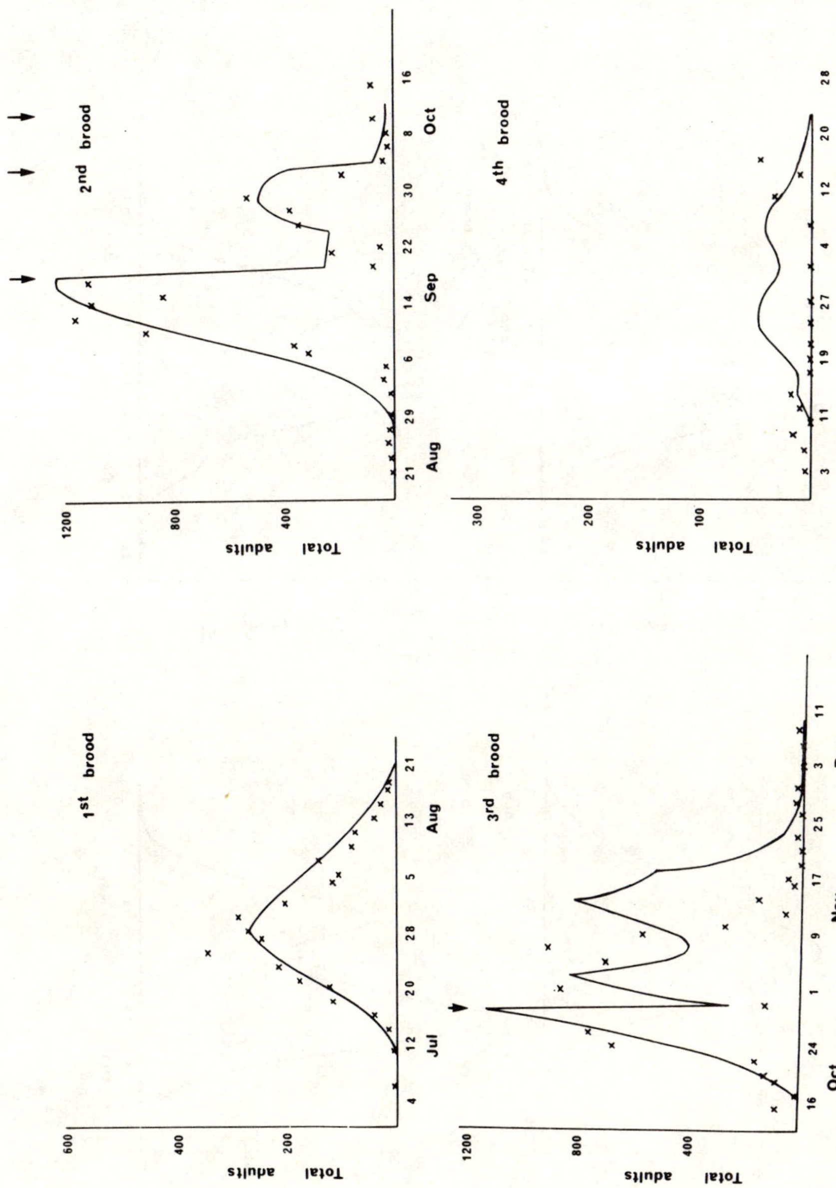

Figure 4. Final fit of the model to adult in the four broods recorded in Orange Field 3 in 1973-74. Arrows indicate days of spraying with Unden.

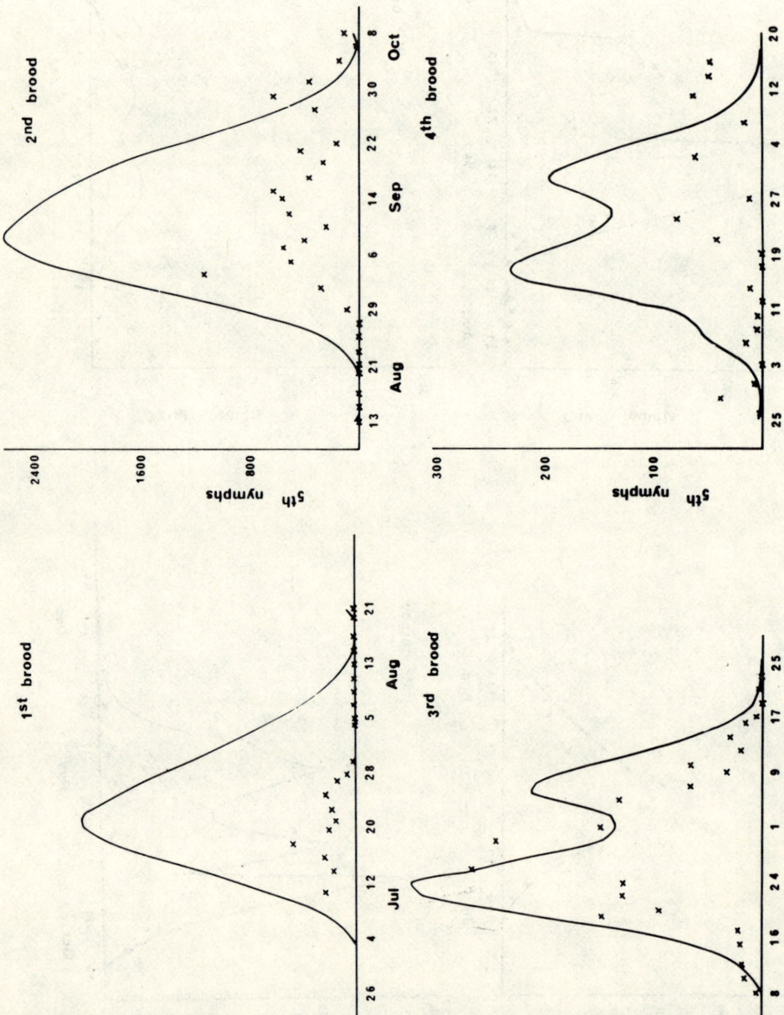

Figure 5. Final fit of the model to fifth instar nymphs in the four broods recorded in Orange Field 3 in 1973-74.

The fit of the model could be improved by adding further components. Nevertheless, at this stage the model fitting procedure suggests a number of conclusions and hypotheses. First, it appears that the sampling procedure considerably underestimates the numbers of nymphs in the first two broods, and monitoring of the growth of the pest population based on sampling this stage is likely to be unreliable, unless the degree of underestimation can be accurately quantified. Second, the daily survival rates of nymphs obtained by fitting the model tended to be higher than expected. They were often as high as, or higher than, the laboratory estimates, whereas it was expected that they would be lower, because of mortalities due to parasites and predators which were excluded from the laboratory experiment. Survival rates were lower however for the first, third and fourth brood of adults, reflecting the presence in the field of either adult migration or mortalities due to predators or disease.

These conclusions are of interest and of some relevance for control but we have found, at least for this pest, that the measurement of the important parameters in the manner prescribed by the Leslie Matrix formulation presents considerable difficulties. We have found that using a model of this kind does not readily stimulate further field work so the feedback between modelling and data acquisition is not greatly encouraged. It is partly for these reasons that Birley (1977) is exploring the value of a transfer function model in tackling this problem.

We have used the Leslie Matrix model of the froghopper to gain some preliminary ideas concerning control. We have, for example, explored the possibility of continuous sterilization of the first brood male population as a means of control. We assumed random mating and complete mixing of the population, and then examined the effects of the proportion sterilization on the subsequent number of adult-days and diapause eggs produced. In Table 2, these results are compared with those obtained when insecticides are applied according to the standard decision rules. The simulations show, at least for one year and one field, that trapping and sterilizing first brood males at a rate of 40% per day will give control as good as current spraying practice. However, a significant amount of damage is still likely to occur, partly due to the fact that the number of adult-days in the first brood is unaffected. A more efficient strategy may be to spray the first brood prior to sterilization.

With the Leslie Matrix model in this form, finding an optimal spraying strategy is a relatively simple problem. The model assumes that survival and fecundity rates change for each brood but in a manner which is independent of the size of the brood. Thus the most effective control measures are those taken against the first brood; the greater the kill in the first brood the lower the population in later broods. This will not hold true, however, if fecundity and survival are density-dependent processes.

Table 2. Simulation results of continuous sterilization of first brood males compared with two spraying strategies, using the model of the froghopper population in Orange Field 3. (Under the decision rule, spraying is carried out when first brood populations exceed 5 adults per 100 stems and when later broods exceed 50 adults per 100 stems.)

Proportion of Fertile Males Sterilized Each Day	Total Adult-Days	Total Diapause Eggs Laid
0	80 088	201 123
0.1	60 058	146 945
0.2	46 225	109 723
0.3	36 139	82 685
0.4	28 480	62 204
0.5	22 474	46 174
0.6	17 643	33 298
0.7	13 674	22 732
0.8	10 357	13 906
0.9	7 543	6 424
1.0	5 126	0
Spraying according to decision rule	27 030	68 820
Spraying as carried out in field	43 332	90 925

To examine what effect density-dependence relationships would have on the optimal spraying strategy, we have used a simpler model of the froghopper population. Using data for a single brood, we have fitted a beta function to give the total number of adults (a_t) at time t :

$$a_t = A\,(3.553 \times 10^{-7})\; t^2\; (55-t)^8 \tag{5}$$

and we have assumed that the total number of adult-days in one generation is related to the number of adult-days in a previous generation by the following relationship:

$$A_{g+1} = \lambda A_g^{\,1-b} \tag{6}$$

where A_g, A_{g+1} = the number of adult-days in generation g and g+1, and λ = the net reproduction rate.

This is the simple density-dependent relationship of Morris (1959) and Varley and Gradwell (1960) whose properties are described in detail by May et al. (1974).

The parameter b measures the degree of density-dependence between the broods. When b = 0, there is no modifying effect of density on the relationship. When $0 < b < 1$, the density-dependence relationship is under-compensating and when $1 < b < 2$, it is over-compensating. Perfect density-dependent regulation occurs when b = 1.

If present in the froghopper life-cycle, density-dependence could affect one or more of several processes relating one brood to the next (Figure 6). Unfortunately estimation of the b value for these processes requires extensive data sets. The data so far available suggests that both nymphal survival and adult migration may be density-dependent.

We have assumed that the adult daily survival rate s is constant for the brood and thus the number of adult-days removed by a single non-residual insecticide application is

$$\frac{a_t\, ks}{1 - s} \tag{7}$$

where k = the proportionate kill
and t = the day on which the spray is applied.

For a residual insecticide, the number of adult-days removed by a single application will be

$$(a_t + \sum_{i=t+1}^{\Gamma} e_i)\; k\; s \tag{8}$$

where Γ = the number of days of residual action
e_i = the number of adults emerging on day i.

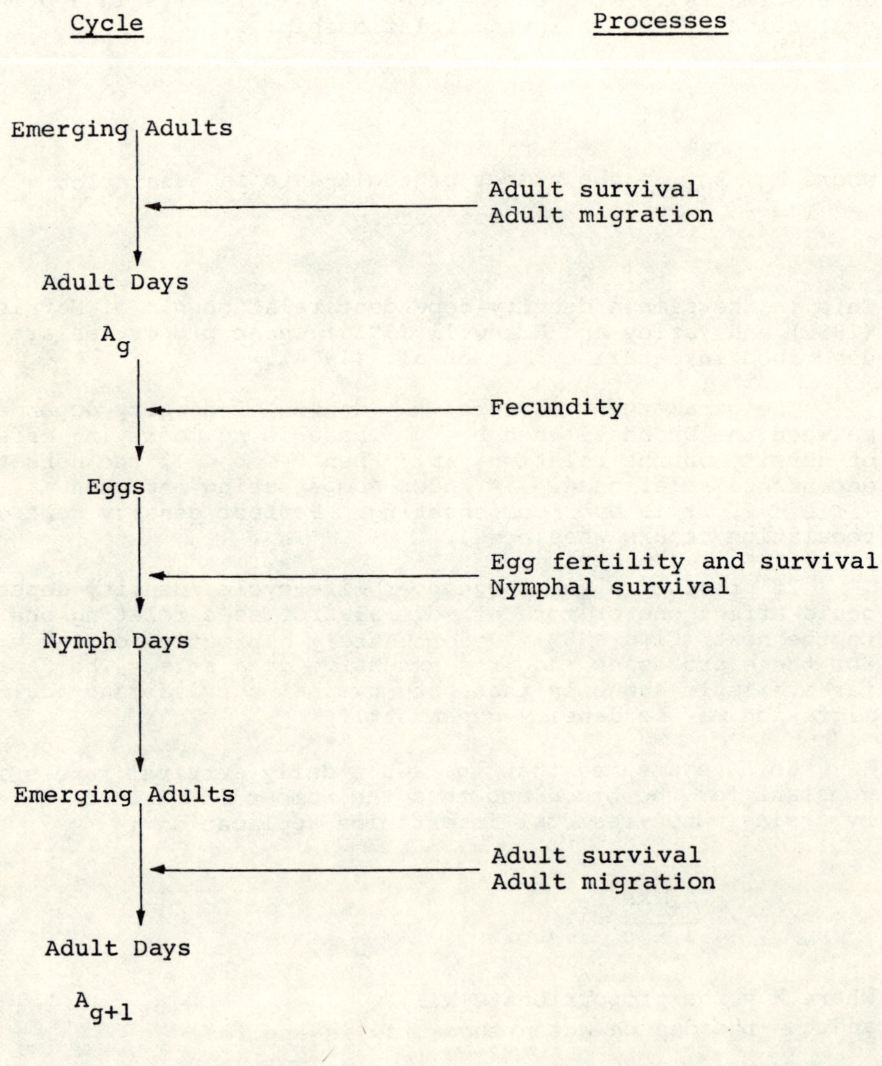

Figure 6. Processes in the life cycle of the sugar cane froghopper liable to be affected by density dependence.

We further assumed that a maximum of three sprays could be applied in a brood using any combination of residual or non-residual insecticide. Table 3 shows the nine dominant strategies for single broods and the optimal times of application.

Table 3. Optimal application times for ten spraying strategies directed against a single froghopper brood

Strat-egy	Time from First Adult Emergence (Days)																		Percentage of Adult-Days Removed
	1	2	3	4	5	6	7	8	9	10	11	12	13	14	15	16	17	18	
1																			0.0
2											NR								19.3
3									NR					NR					32.7
4										R									37.6
5							NR				NR					NR			42.8
6									NR			R							49.0
7							NR			NR			R						57.2
8							R							R					60.9
9						NR		R							R				67.4
10				R						R						R			74.9

R = residual spray; NR = non-residual spray.

A strategy s is profitable when

$$Pd\alpha(S) > C(S) \qquad (9)$$

where $\alpha(S)$ = the number of adult-days eliminated by strategy S. This condition is essentially equivalent to the economic threshold criteria of Stern et al. (1959). For a non-residual insecticide (expression (7)) the condition can be expressed as

$$a_t > C(NR)\,\frac{1-S}{PdkS} \qquad (10)$$

and for a residual insecticide (expression 8), as

$$a_t + \sum_{i=t+1}^{\Gamma} e_i > C(R)\,\frac{1-s}{Pdks} \quad . \qquad (11)$$

Both of these criteria are practicable where a_t and e_t are being monitored and only a single spray is contemplated. But where more than one spray is to be applied, a systematic computer search procedure has to be followed. Figure 7 shows the result for a single brood. Since the damage function is linear the net revenue curves are also linear. A strategy is profitable when its net revenue line intersects the non-sprayed revenue line (strategy 1). For each brood size the uppermost revenue line indicates the most profitable strategy and thus the arc subtended by these upper lines describes the optimal spraying policy for a range of brood sizes.

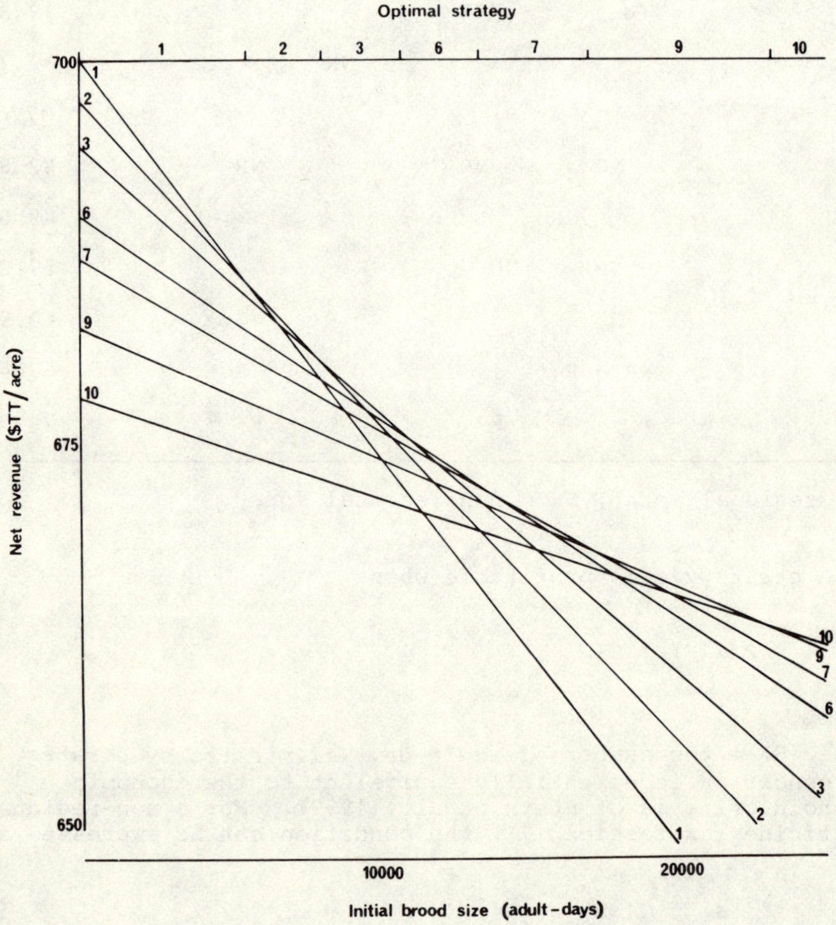

Figure 7. Optimal spraying strategies (see Table 3) against a single froghopper brood.

Extending the analysis to the four brood condition which occurs in the field is computationally laborious. The solution is most efficiently found by the application of the dynamic programming technique (Bellman, 1957, Shoemaker 1973, Watt, 1963). Table 4 shows the optimal spraying strategy obtained by this technique for three different assumptions about density-dependence. The result clearly illustrates the importance of density-dependence for control policy. Heavy first brood spraying is only optimal when there is little or no density-dependence; the spraying shifts to the second and third broods as the degree of density-dependence increases.

Table 4. Optimal spraying strategies against all four froghopper broods with varying degrees of density-dependence. Strategies as in Table 1 (after Conway et al. 1975)

Initial Size of 1st Brood Adult-Days	Density-dependence b=0 (None) BROOD				b=0.5 (Undercompensating) BROOD				b=1.25 (Overcompensating) BROOD			
	1st	2nd	3rd	4th	1st	2nd	3rd	4th	1st	2nd	3rd	4th
2000	7	3	1	1	1	10	3	1	1	10	10	1
5000	10	5	1	1	6	10	3	1	1	9	10	1
10000	10	10	1	1	10	10	3	1	2	9	10	1

These optimal spraying policies have also been examined by Comins (1977) who has assessed the effect of possible pesticide resistance in the froghopper. Table 5 shows the costs of the strategies both as computed above and taking into account possible resistance. In the presence of undercompensating density-dependence, large brood numbers result in greatly increased resistance costs. But in the presence of overcompensating density-dependence, the resistance costs remain more or less constant over a wide range of initial brood sizes.

Table 5. Effect of resistance on the cost of optimal spraying strategies for the froghopper (after Comins 1977)

Density-Dependence	Initial Size of 1st Brood (Adult-Days)	Total Cost ($) No Resistance	Total Cost ($) Resistance
	2000	59	59
b=0.5	5000	73	77
	10000	84	99
	2000	90	98
b=1.25	5000	95	100
	10000	104	109
	2000	35	35
b=0	5000	49	53
	10000	60	75

THE CATTLE TICK (BOOPHILUS MICROPLUS)

The cattle tick is the most important pest of cattle in Australia and is estimated to cost the industry A$40 million per annum in control costs and production losses. Damage to cattle is caused by adult female ticks, their feeding resulting in loss in liveweight gain as well as damage to hides. The engorged females then detach themselves from the host and lay eggs in the pasture. These hatch into free living larvae which are in turn picked up by the grazing cattle. The duration of the parasitic phase of the life-cycle is a constant three weeks while the non-parasitic phase may vary in duration from four to twenty weeks depending on the season and availability of hosts.

In Queensland the most widely used method of control is to dip the cattle in a chemical acaricide to kill the parasitic ticks on the cattle, but as Table 6 shows this has resulted in very rapid evolution of acaricide resistant strains (Wharton and Roulston, 1970). Other methods of tick control are possible.

Table 6. Development of acaricide resistance in the Australian cattle tick (courtesy of W.J. Roulston & R.H. Wharton)

Chemical	Year Introduced	Year in Which Resistance Widespread
Arsenic	1895	1936
DDT	1946	1955
BHC	1950	1952
Diazinon	1956	1963
Dioxathion	1958	1963
Carbaryl	1963	1963
Carbiphenothion	1961	1963
Coumaphos	1959	1966
Ethion	1962	1966
Chlorpyrifos	1967	1970
Bromophosethyl	1967	1970

For example, the technique of pasture spelling, in which cattle are removed from the pasture for a period of time, produces increased mortality of free living larvae. Zebu type cattle (Bos indica L) are also more resistant to tick attack and use of this breed or cross-bred European cattle (Bos taurus L.) provides another form of control (Wharton, Utech and Turner 1970; Wharton, Utech and Sutherst 1973). However, these methods have their own disadvantages and neither is an exclusive alternative to dipping. A possible solution to the resistance problem is to combine the three methods, exploiting the advantages of each.

The aim of our case study has not been to design an optimal control strategy for any particular property or region but to develop a framework and provide a set of models and techniques which will aid in the development of improved strategies. We have so far only examined the dipping strategies. To do this we have used a modified Leslie Matrix model for the cattle tick, and divided the population structure into weekly age classes, thus giving 13 developmental classes:

$n(0)_t - n(4)_t$	eggs)	) in pasture
$n(5)_t - n(8)_t$	larvae)	
$n(9)_t - n(11)_t$	parasitic ticks on cattle	
$n(12)_t$	adult females in pasture	

so that the Leslie Matrix is given by

$$
\begin{pmatrix}
\cdot & \cdot & \cdot & \cdot & \cdot & \cdot & \cdot & \cdot & \cdot & \cdot & \cdot & \cdot & \cdot \\
\cdot & \cdot & \cdot & \cdot & \cdot & \cdot & \cdot & \cdot & \cdot & \cdot & \cdot & \cdot & \cdot \\
\cdot & e & \cdot & \cdot & \cdot & \cdot & \cdot & \cdot & \cdot & \cdot & \cdot & \cdot & \cdot \\
\cdot & \cdot & e & \cdot & \cdot & \cdot & \cdot & \cdot & \cdot & \cdot & \cdot & \cdot & \cdot \\
\cdot & \cdot & \cdot & e(1-h_3) & e(1-h_4) & \cdot & \cdot & \cdot & \cdot & \cdot & \cdot & \cdot & \cdot \\
\cdot & \cdot & \cdot & eh_3 & eh_4 & \cdot & \cdot & \cdot & \cdot & \cdot & \cdot & \cdot & \cdot \\
\cdot & \cdot & \cdot & \cdot & \cdot & \ell(1-p) & \cdot & \cdot & \cdot & \cdot & \cdot & \cdot & \cdot \\
\cdot & \cdot & \cdot & \cdot & \cdot & \cdot & \ell(1-p) & \cdot & \cdot & \cdot & \cdot & \cdot & \cdot \\
\cdot & \cdot & \cdot & \cdot & \cdot & \cdot & \cdot & \ell(1-p) & \ell(1-p) & \cdot & \cdot & \cdot & \cdot \\
\cdot & \cdot & \cdot & \cdot & \cdot & \ell p & \ell p & \ell p & \ell p & \cdot & \cdot & \cdot & \cdot \\
\cdot & \cdot & \cdot & \cdot & \cdot & \cdot & \cdot & \cdot & \cdot & (1-k) & \cdot & \cdot & \cdot \\
\cdot & \cdot & \cdot & \cdot & \cdot & \cdot & \cdot & \cdot & \cdot & \cdot & (1-k) & \cdot & \cdot \\
\cdot & \cdot & \cdot & \cdot & \cdot & \cdot & \cdot & \cdot & \cdot & \cdot & \cdot & 1 & \cdot
\end{pmatrix}
$$

where e = constant egg survival rate
h = proportion of eggs hatching, in week 3 (h_3) and thereafter (h_4)
l = constant larvae survival rate
p = proportion of larvae taken up by cattle per week
k = constant mortality rate of parasitic stages due to acaricide dipping.

This basic matrix is then modified to take account of the density-dependent mortality which is present in the early parasitic stage. Sutherst et al. (1973) have demonstrated that competition between parasitic ticks leads to a relationship of the form:

$$M = a\,L^{1-b}$$

where M = the number of females reaching maturity,
L = the number of larvae picked up, and
a,b = density-dependence coefficients.

We have assumed that this relationship operates within the first week of parasitism so that

$$n(9)_t = \alpha[2\ell p(n5)_t + n(6)_t + n(7)_t + n(8)_t)]^{1-b}$$

where ℓ = larval weekly survival rate
p = weekly host finding rate

Parameter values for the model were derived from published and unpublished data, wherever possible using data relevant to South East Queensland conditions. It was assumed that values changed between but not within broods and the values chosen were judged to be typical of average climatic conditions in South East Queensland. When run with these values the model gave a fairly realistic pattern of numbers for a single year (Figure 8). Note that the populations on Zebu cattle are very much lower than on European cattle because of the much greater degree of density-dependent competition which occurs on Zebus.

We have decided not to attempt further adjustments of the model to improve its realism. A more complex and sophisticated model is being developed elsewhere. We feel that the model presented here is sufficiently realistic to explore at least the general implications of various control strategies. Using the model we have thus investigated the importance of the relationships between generations and the pattern within generations for dipping strategies.

Dipping cattle in an acaricide kills a large proportion of all ages of parasitic ticks. Since it takes approximately three weeks from the time larvae attach to cattle for them to become adult females, a single dipping produces a three week period when the treated cattle are relatively free of adult ticks.

Figure 9 shows the total number of tick weeks resulting from a single dip in each generation for a wide range of initial egg densities. It is clear from these results that dipping the first generation is considerably more effective than dipping later generations. The figure also illustrates the effect of density-dependence. In the absence of density-dependence the relationship between total tick weeks and initial egg density would be linear rather than curved, as here. In the case of European cattle, Figure 9, the degree of density-dependence is light and the relationship is only slightly non-linear, whereas for Zebu cattle (not shown here) the curvature is more pronounced due to stronger density-dependence.

It is interesting to note that if we ignore the age class effects (associated with the Leslie Matrix model) and simply assume that the generations are linked by a simple density-dependent relationship of the form

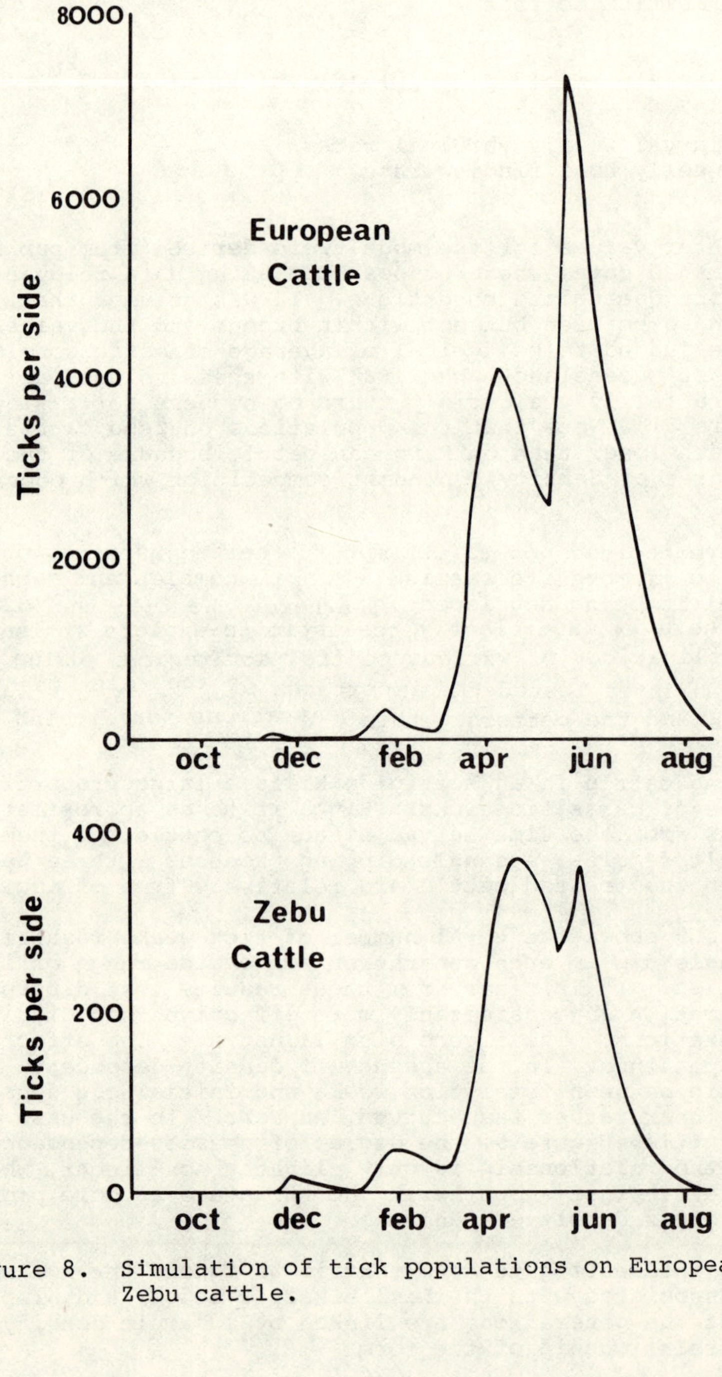

Figure 8. Simulation of tick populations on European and Zebu cattle.

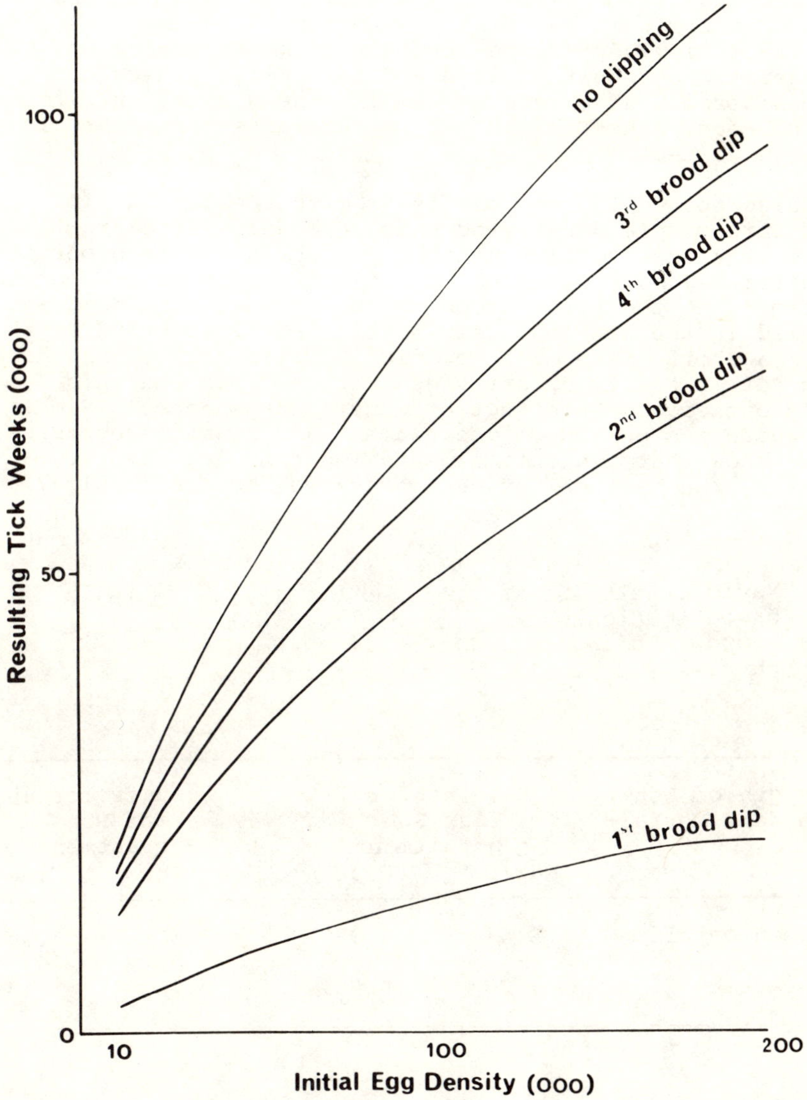

Figure 9. Effect of dipping different broods on European cattle (99% kill).

$$N_{g+1} = \lambda N_g^{1-b}$$

where g refers to generations 1,2,3 and 4 and g+1 is 2,3,4, and 1 respectively, it can be shown that in the case of under-compensating density-dependence, as occurs here, i.e. $0 < b < 1$, a single dipping is more effective when applied against the third brood rather than the first. The reason is that the presence of density-dependence attenuates the reduction in future generations resulting from a single kill. There is thus a considerable advantage in dipping the generation before the fourth brood, since the latter is the largest and the dip has a greater effect.

The discrepancy in the results from the Leslie Matrix model and this simple density-dependence model arises from the importance of the generation pattern, which is not incorporated in the latter. Its importance lies in the length of the period of recruitment to the first brood of the season. If, as has been assumed in the Leslie Matrix model, the first brood eggs hatch over a short space of time, dipping will result in a high proportion of mortality of early generation ticks and this apparently outweighs the effect of density-dependence. Simulations in which the period of recruitment and density-dependence are varied show that the optimal spraying time depends very critically on the balance between these two factors (Table 7).

Table 7. Ratio of tick weeks produced by dipping in third brood to tick weeks produced by dipping in first brood. Spring recruitment is assumed to be narrowly or widely distributed in time.

The Strength of Density-Dependent Mortality	Results for 1 Year Run Recruitment		Results for 10 Year Run Recruitment	
	Narrow	Wide	Narrow	Wide
European cattle	2.01	0.98	7.50	1.34
Zebu cattle	1.24	0.95	2.14	1.13

We further used the Leslie Matrix model to examine the effect of more than two dips in the season, and found that when three or more dips are applied, the resulting tick weeks, and hence damage, are largely independent of initial egg density. We also found that the best strategies are only slightly affected by changes in the development rate of the tick population. In general it would thus seem that the best strategy arrived at by this procedure is fairly robust, being only drastically altered by changes in the pattern of egg hatch at the beginning of the season or the degree of density-dependence between generations.

REFERENCES

Bellman, R. (1957), *Dynamic Programming*, Princeton University Press, N.J.

Birley, M. (1977), A transfer function, pest management model (this volume).

Comins, H.N. (1977), The management of pesticide resistance, *J. Theor. Biol.* (in press).

Conway, G.R., G.A. Norton, A.B.S. King, and N.J. Small (1975), A systems approach to the control of the sugar cane frog-hopper, in *Study of Agricultural Systems*, G.E. Dalton, ed., Applied Science Publishers, London, 193-229.

Leslie, P.H. (1945), On the use of matrices in certain population mathematics, *Biometrika*, 33, 183-212.

May, R.M., G.R. Conway, M.P. Hassell, and T.R.E. Southwood (1974), Time delays, density dependence and single species oscillations, *J. anim Ecol.*, 43, 747-770.

Morris, R.F. (1959), Single factor analysis in population dynamics, *Ecology*, 40, 580-588.

Shoemaker, C. (1973), Optimization of agricultural pest management II, Formulation of a control model, *Math. Biosciences*, 17, 357-365.

Southwood, T.R.E. and G.A. Norton (1973), Economic aspects of pest management strategies and decisions, in *Insects: Studies in Population Management*, P.W. Geier, L.R. Clarke, D.J. Anderson and H.A. Nix (eds.), Ecol. Soc. Aust Mem. 1 Canberra, 168-184.

Stern, V.M., R.F. Smith, R. Van den Bosch, and K.S. Hagen (1959), The integrated control concept, *Hilgardia*, 29, 81-101.

Sutherst, R.W., K.B.W. Utech, M.J. Dallwitz and J.D. Kerr (1973), Intra-specific competition of Boophilus microplus (Canestrini) on cattle, *J. Appl. Ecol.*, 10, 855-862.

Sutherst, R.W., G.A. Norton, N.D. Barlow, G.R. Conway, M. Birley, and H.N. Comins, Analysis of management strategies for the cattle tick (Boophilus microplus) in Australia (in preparation).

Usher, M.B. (1972), Developments in the Leslie Matrix model, in *Mathematical Models in Ecology*, J.N.R. Jeffers, ed., Symp. Brit. Ecol. Soc., Blackwells, Oxford, 29-60.

Varley, G.C. and G.R. Gradwell (1960), Key factors in population studies, *J. Anim. Ecol.*, 29, 399-401.

Watt, K.E.F. (1963), Dynamic programming, "look ahead programming" and the strategy of insect pest control, *Canad. Ent.*, 95, 525-636.

Wharton, R.H., K.B.W. Utech, and R.W. Sutherst (1973), Tick resistant cattle for the control of Boophilus microplus, *Proc. 3rd Int. Cong. Acrud.*, Prague, 697-700.

Wharton, R.H., K.B.W. Utech, and H.G. Turner (1970), Resistance to the cattle tick, Boophilus microplus in a herd of Australian Illawarra Shorthorn Cattle: its assessment and heritability, *Aust. J. Agric. Research*, 21, 163-181.

A TRANSFER FUNCTION, PEST MANAGEMENT MODEL

Martin Birley

INTRODUCTION

One of the principal challenges to the theory of insect pest management is to describe the interactions of populations with age structure in a seasonal environment (Sinko and Streiffer, 1967; Oster and Takahashi, 1974; Auslander, Oster and Huffaker, 1974). An ideal technique would be both realistic and easily communicable, while containing mathematically tractable non-linear equations. It would mimic salient features of population dynamics and enable alternative management strategies to be explored. In the absence of such a chimera, some improvization is necessary which fulfills a more limited goal. Non-linear partial differential equations (op. cit.) represent one solution to this problem.

In the face of complexity, the simple, classical techniques of linear systems theory may have been overlooked. The transfer function or convolution, in discrete or continuous form, is at the core of this theory, is well known in the bio-engineering literature (Milsum, 1966), and has been mentioned by ecologists (Kowal, 1971; Hubbell, 1971). Indeed, the basic age-specific survival and fecundity processes of populations are convolutions, and were described as such by demographers such as Lotka (1939), Rhodes (1940) and Leslie (1948). Transfer functions have been used to model populations by Lewis (1972, 1976) and to forecast population change by Hacker et al. (1975).

Before transfer function techniques can be applied, the population process must be decomposed into linear and non-linear components. A transfer function model of some of the principal linear components may describe the major features of population change. When non-linear components are added the elegant properties of the linear theory are lost but the technique remains conceptually simple and may be simulated on a computer.

It may not be possible to apply a single mathematical technique to all insect pest modelling problems. However, it may be possible to group pest management problems into categories, or paradigms, according to the techniques which are most appropriate for their description. The assumptions which have been made about the problem must be explicitly stated and constitute the paradigm definition. One such paradigm concerns multi-voltine pests with discrete generations in a seasonal environment. A case study of this paradigm is the sugar cane froghopper (Aeneolamia varia saccharina, Dist., Homoptera, Cercopidae) which is an important pest of sugar cane in Trinidad (Fewkes, 1969).

THE CONTROL PROBLEM

Components abstracted from the pest management problem help to delineate the paradigm. One such component is the set of admissable controls. I restrict this set to the timing of pesticide application, and assume that a non-residual pesticide spray of known cost and kill rate can be applied on any day in the growing season. The control problem is to choose the spraying dates.

Another component of the control problem is the precise relationship between crop yield, pest damage and the monetary value of the crop. This relationship can be complicated by plant compensation, asynchrony between plant and pest growth rate, qualitative effects of crop damage and many other factors (Southwood and Norton, 1973). I assume that damage is restricted to a single instar and is linearly related to total instar-days throughout the course of the growing season. I also assume that the monetary value of the crop is linearly related to damage. By making these assumptions, the details of plant growth can be ignored and attention can be restricted to the population dynamics of the pest. Although the problem becomes less realistic it becomes much more amenable to theoretical study.

The objective of pest management is assumed to be an increase in the net profit obtained from a single crop. Therefore, the objective function is to maximize:

monetary value of the crop - monetary cost of pest control.

But since the monetary value of the crop is assumed to be linearly related to total instar-days, there is an equivalent and simpler objective function which is to minimize:

monetary cost of damage caused by total instar days + monetary cost of pest control.

It is assumed that each application of pesticide increases the monetary cost of pest control but decreases the monetary cost of total instar-days of pest damage. The economic threshold can now be defined as that level of pest infestation at which the cost of control action is just equal to the cost of damage which would be prevented by that control action.

By examining the components of the control problem, attention is focussed on a particular instar within the pest life-cycle, on the number of individuals within that instar throughout the entire season, and on the effect of control action on the number of individuals. A dynamic model is required to relate these factors. The effect of pesticide timing can then be investigated on the model and some insight gained as to the effect of timing on the real pest system.

THE PEST LIFE-CYCLE

The pest population has a seasonal diapause which destroys the population age structure and distinguishes the paradigm from, for example, human populations, where age structure has continuity over time. At the beginning of each growing season the population is recruited from the diapause state, and this determines both the age structure and the pest infestation rate. Development proceeds through several instars, development rate apparently being independent of daily fluctuations in environmental conditions. This assumption distinguishes the paradigm from those pest species where development rate is critically dependent on climate.

The most important assumption that I wish to include in the paradigm is that the probability of survival is age-specific. After recruitment to a particular instar, some proportion of a cohort survives each age interval. I assume that a large component of daily mortality is dependent on the age of the individual. There will certainly be age-independent sources of mortality, such as pesticide application or climatic extremes, and these will be referred to as catastrophes. Catastrophic mortality is assumed negligible unless mentioned explicitly. I also assume that the fecundity of female adults is age-specific. It is the age-specific nature of the paradigm which makes the transfer function approach most appropriate. Finally, I assume that the instar which causes economic damage is the same as the instar which can be controlled by pesticide application. Consequently, attention can be focussed on recruitment to a specific instar and the age-specific survival of recruits. If the pest is multivoltine, the surviving individuals will produce offspring which ultimately appear again as recruits to the damaging instar at a later time within the season. In this case, there is an inter-generational relationship which determines how many offspring recruits to an instar are produced per instar recruit.

THE CASE STUDY

The sugar cane froghopper overwinters as eggs, which are hidden in the soil around the base of the ratooned, sugar cane crop. Hatchling recruits begin to emerge at the beginning of the growing season, in response to soil moisture, and the nymphs attach to the plant roots, surround themselves with spittle, and remain there until the last nymphal moult. They feed on xylem sap but do not cause appreciable damage. The last nymphal moult takes place in a special froth chamber, situated at the soil surface. The adults live in the upper leaf axils of the sugar cane plants and their feeding punctures cause necrosis, or blighting, of the leaves. The loss of photosynthetic tissue retards plant growth and reduces the yield of sugar when the cane is harvested at the end of the season. Damage is controlled by aerial spraying of insecticide which is directed against adults. There is a scouting procedure to determine the best time to spray. An additional point of importance in the following discussion is that adult recruitment may be monitored by sampling the froth chambers occupied by pre-emergent adults. There are four generations, or broods, in the season, although the last brood is small since most of the eggs produced by the third generation enter the diapause state.

It should be pointed out that the pest could probably be controlled by frequent ploughing (Fewkes 1969) but, for economic reasons, it is preferable to ratoon the crop for as many seasons as possible. Consequently, pest infestation builds up, pesticides must be used, and the environment may be damaged. However, the only pest management issue which is under discussion here is the optimal timing of pesticide application, although by pursuing an optimal policy, it may be possible to minimize pesticide application and, hence, environmental damage.

THE MODEL CHOICE

The properties of the control problem and the population life-cycle together restrict model choice. For example, a model must be chosen which includes age classes, instars and recruitment. But the problem of parameter estimation should be considered before a model is finally selected. The values of the model parameters will have to be adjusted so that the model output corresponds to a set of, generally, scarce and noisy observations which have been made of the pest system.

Insect population samples are often estimates of instar density obtained at several day intervals throughout the season. It is of importance that the data is grouped by instar rather than by age class. The parameters of an age class model, such as the Leslie Matrix (Lewis, 1942; Leslie, 1945; Usher 1972), cannot be estimated directly from time series of instar density data (Lefkovitch, 1965), A matrix model based on instar class rather than age class will only work if the age structure is stable (Vandermeer, 1975).

THE TRANSFER FUNCTION

A transfer function model has several advantages over a matrix model for describing the paradigm of an insect population in a seasonal environment. A recruitment time series is easily incorporated as the transfer function input, and instar density, rather than age class density, is a natural choice for the transfer function output. The transfer function model is already an underlying feature of many methods of analyzing insect population data and estimating population parameters (Manly, 1974; Birley, 1977a). Therefore, a transfer function is an appropriate model for both the control problem and the estimation problem. Figure 1 illustrates this relationship.

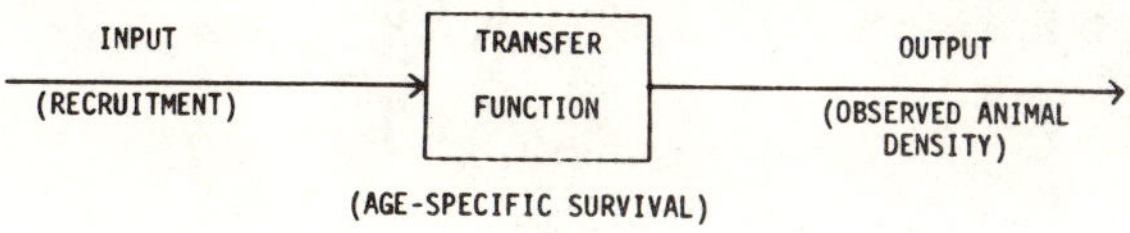

ESTIMATION

Given the input and output time series, determine the transfer function.

PREDICTION/SIMULATION

Given the transfer function determine:

1. the effect of varying the input;
2. a prediction of the output;
3. the effect of a control strategy;
4. how to link different stages and/or species;
5. general mathematical properties of the system.

Figure 1. The relationship between estimation and control theory.

There are many different transfer function models and the discrete form, using a time base of one day, is appropriate for this particular paradigm. Figure 2 illustrates the model. The input (f(t)) is a time series of instar recruitment density, the output (N(t)) is a time series of observations or predictions of instar density and the transfer function is a set of parameters (S_j) which represent the age-specific survivorship function, i.e. the probability of surviving j days. The set of constraints attached to the parameter values embody the principle that animals must die as they age.

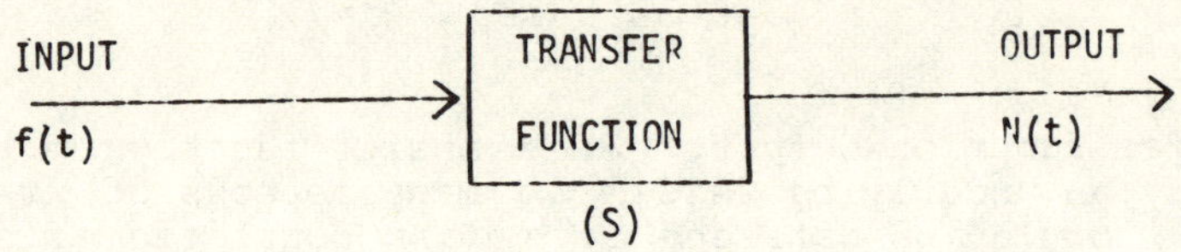

$$\text{MODEL:} \quad N(t) = \sum_{j=0}^{d} f(t-j).S_j$$

$$\text{CONSTRAINT:} S_o \geqslant S_1 \geqslant S_2 \geqslant \ldots \geqslant S_d \geqslant 0$$

Figure 2. The transfer function model.

I shall use some mathematical shorthand in the following discussion. This shorthand is illustrated in Figure 3 and replaces the summation sign equation by an operator Φ. The symbol Ø is also used and denotes the backwards operator or unit time delay.

MATHEMATICAL NOTATION USING THE BACKWARDS OPERATOR

Let Ø be defined as the backwards operator which has the property:

$$\emptyset^j.f(t) = f(t-j)$$

The transfer function equation may then be expressed as follows:

$$N(t) = \sum_j S_j.f(t-j)$$

$$= (\sum_j S_j.\emptyset^j).\ f(t)$$

Let Φ(S) be shorthand notation for the polynomial in Ø and S.

$$N(t) = \Phi(S).f(t)$$

Figure 3. Some mathematical shorthand.

TWO SOLUTIONS TO THE CONTROL PROBLEM

First, I consider the case of a uni-voltine pest, which is in agreement with the paradigm outlined. The problem of pesticide timing for this pest can be divided into two separate problems, depending on the extent of recruitment information available. If recruitment can be monitored but not predicted then there is an economic threshold control problem. If recruitment can be predicted in advance for the whole of the season then there is an optimal control problem. By using threshold control, it should be possible to improve profits and, by using optimal control, it should be possible to maximize profits. But in order to maximize profits it is necessary to have the additional information about future pest recruitment.

For the threshold control problem it is assumed that recruitment has been monitored up to, and including, each decision day. It would then be rational to spray if the reduction in damage associated with spraying is greater than the cost of spraying. It is assumed that damage can be measured in instar-days and therefore it is necessary to know the number of instar-days which could be eliminated by spraying on a given day. If $c(t)$ is the fraction of survivors from the non-residual spray, $Q(t)$ is the cost of the spray and $e(t)$ is the monetary damage value of one instar-day, then the economic threshold (Stern et al., 1959; Norton, 1976) is defined by:

$$D(t) \geq Q(t)/(1-c(t))\cdot e(t) \ . \tag{1}$$

The controllable instar-days $(D(t))$ can be determined from the recruitment time series $(f(t))$ and the survivorship function (S) in a similar manner to the instar density $(N(t))$. The transfer function relationship between $N(t)$ and $f(t)$ was illustrated in Figure 2. The relationship between $D(t)$ and $f(t)$ depends on the parameter α which is defined by:

$$\alpha_j = \sum_{i=j}^{d} S_i \tag{2}$$

where d is the stage duration, α_j determines the number of instar-days which will be contributed by an individual of age j. Then, using the notation of Figure 3, it follows that:

$$D(t) = \Phi(\alpha)\cdot f(t) \ . \tag{3}$$

These arguments are summarized in Figure 4. The recruitment is known up to and including day t (M is an unimportant parameter, of value unity in this case). When recruitment has been monitored both N(t) and D(t) can be estimated, providing that the instar survivorship function is known. If the estimated value of D(t) exceeds the economic threshold, defined by equation (1), it is rational to spray.

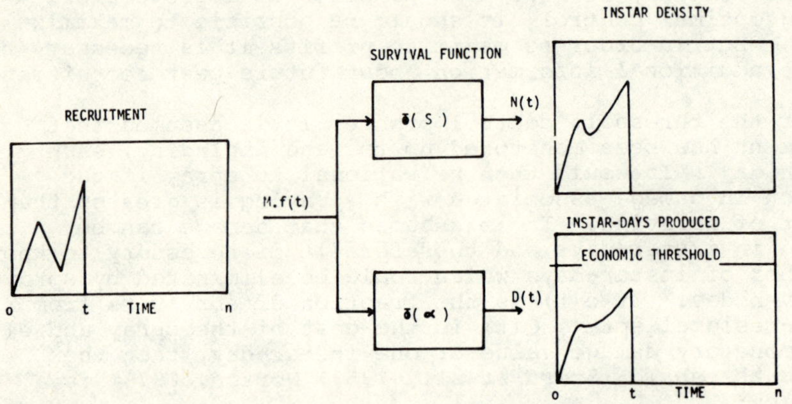

Estimation of the threshold for insecticide application by instar recruitment monitoring.

Figure 4. A model for determining threshold spraying rules from recruitment data.

In order to determine the optimal control strategy for a uni-voltine pest, the entire sequence of recruitment (f(t)) must be known (t=0,n). The optimal control minimizes

$$FS = \sum_{t=0}^{n} (e(t) \cdot N(t) + Q(t)) \qquad (4)$$

where N(t) is the instar density on day t subject to the control sequence c(t).

The optimization problem is solved by application of a modified dynamic programming technique. Normal dynamic programming (Bellman, 1957) is based on an autoregressive model, in which state transitions depend on the previous state and the control action. In a moving average model, the state transitions depend on the recruitment time series, the previous control sequence, and the control action. The state (K) at time t depends on the control sequence (c(j), j=t, t-d) where the instar duration is d. If the control action is binary valued (spray or not spray) there are 2^d states to be considered at each time interval. The mathematical details of the technique are described elsewhere (Birley, 1977a). A normal recursive formula is used in which FT(t,K) is the minimum cost of going from state K at time t to the end, and state L is the new state when an optimal decision is made at time t.

$$FT(t,K) = FT(t+1,L) + e(t)\cdot c(t)\cdot N_k(t) + Q(t) \qquad (5)$$

where, $N_k(t)$ is the potential value of N(t) when in state K but before control action c(t).

Figure 5 represents a complete, worked example of no control, threshold control and optimal control. The example includes a simple uni-modal recruitment time series, with a peak recruitment of 80 individuals/day, and a survivorship function, with an instar duration of six days. In effect, the control problem is to minimize the area under the curve, subject to the cost of control. Each "spray" is assumed to cost 100 instar-days and kill 0.8 of the population present.

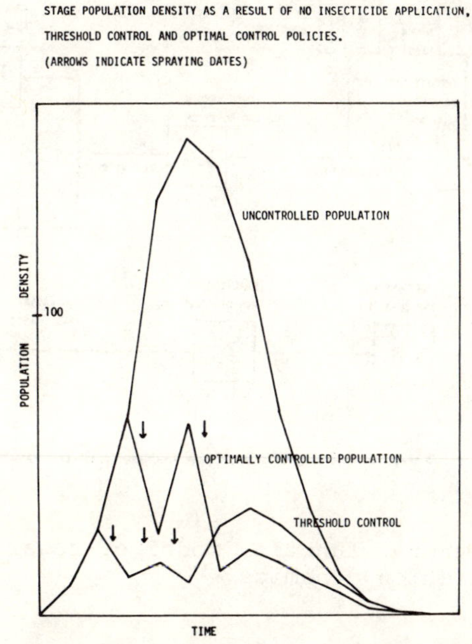

Figure 5. An example of different control strategies.

When the population is not controlled, a total of 792 instar-days of damage is generated. The threshold control policy requires three sprays (cost 300 instar-days) and reduces the damage to 216 instar-days. The total cost of the threshold policy is 216+300 = 516 instar-days, i.e. a saving of 35% over no control. The optimal control policy requires two sprays (cost 200 instar-days) and generates 274 instar-days of damage. The total cost of optimal control is, therefore, 474 instar-days. The net saving must be offset against the cost of forecasting the future, which is necessary in the optimal spraying case.

THE SUGAR CANE FROGHOPPER

In theory, then, the uni-voltine, single instar, pesticide timing problem can be solved. Naturally, when the method is applied to a real, multi-voltine pest, such as the sugar cane froghopper, certain complications arise.

In order to analyze the froghopper problem I had to determine: a transfer function model of the pest life-cycle; estimates of the age-specific adult survival and fecundity; the inter-generational relationship; and the effect of density-dependent processes. Figure 6 is a basic transfer function model of the pest life-cycle. It includes the nymph-adult survival and development rate; the age-specific adult survival and fecundity; egg survival and development; and the hatchling recruitment. It also outlines a possible pathway for hatchling density-dependent mortality which I shall discuss later.

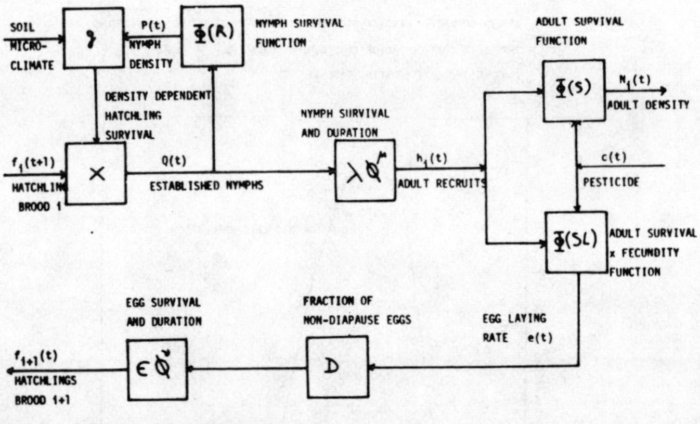

A transfer function model of sugar cane froghopper population dynamics.

Figure 6. A transfer function model of sugar cane froghopper population dynamics.

I obtained estimates of the adult age-specific survival by analyzing observed, adult recruitment and instar density data (Birley, 1977b). Estimates of the inter-generational parameters, especially egg-nymph survival, were obtained by comparing the expected, adult recruitment in successive generations with the observed recruitment, as in Figure 7 (Birley, 1977a). At best, the interbrood model was fairly imprecise and accounted for only 60-70% of the variance. The error could be attributed to numerous random disturbances which must act on natural populations. There were also some sources of error which appeared to be systematic.

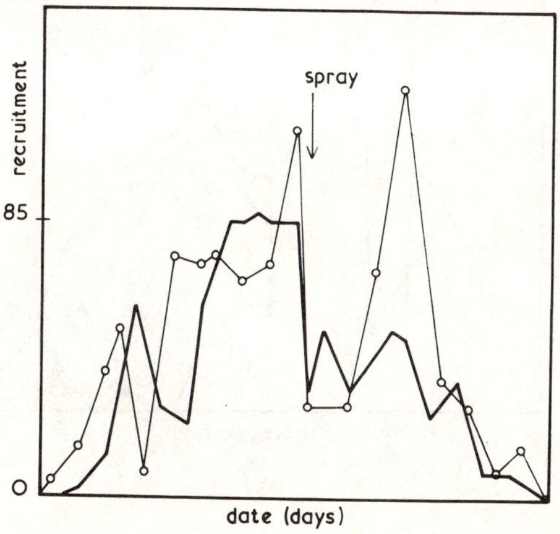

Figure 7. The interbrood relationship for adult froghopper recruits. The observed (o) and expected brood 2 recruitment. (Field OF3.)

Up to this point I have assumed that the model is linear; that is, it obeys the principle of superposition. The control problem for a linear, multi-voltine, pest model is essentially the same as the uni-voltine case. But if there are density-dependent processes operating, the model becomes non-linear and the control problem is much more complex. For example, Conway et al. (1975) demonstrated that, under certain assumptions regarding density-dependence, the optimal time of spraying shifts from the first generation to the third generation.

When I analyzed the relationship between adult froghopper recruitment in consecutive generations, I discovered certain anomalies at high densities which could not be explained by a linear model (see Figure 8). It is equally true, however, that they could not be explained by Conway et al.'s model of density-dependence. These anomalies can be explained by the following hypothesis.

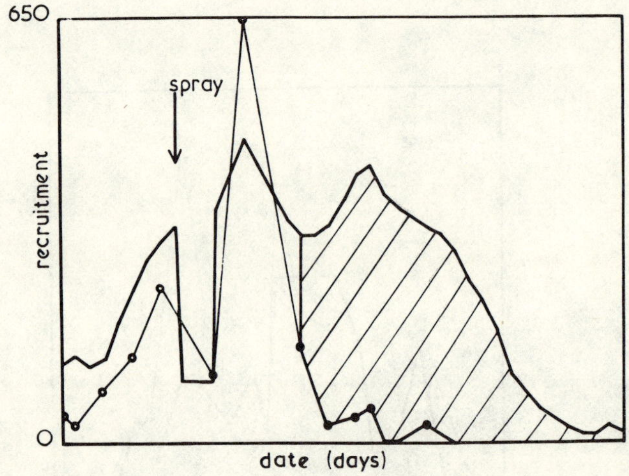

Figure 8. The interbrood relationship for adult froghopper recruits. The observed (o) and expected brood 3 recruitment. (Field OF3).

There is a threshold density below which density-dependent mortality is negligible and a linear model will suffice. Above the threshold, density-dependent mortality increases rapidly with density. The density effect may occur at the hatchling stage, when newly emerged nymphs search for favorable feeding sites which are already occupied by older nymphs. This process can be represented mathematically by a non-linear feedback loop, as in Figure 6. Hatchling survival is modified by a multiplier (g), which is a function of the density of established nymphs (P(t)) and of exogenous variables, such as the state of the soil microclimate.

In this case there are two control problems: a simple control problem, similar to the one I discussed earlier; and a more complex, non-linear problem, which operates above the density effect threshold. The importance of the non-linear problem depends on the relative levels of the economic and density effect thresholds. If the economic threshold is much lower than the density effect threshold, then the control policy should ensure that the density effect threshold is never exceeded and a linear model should suffice. If, on the other hand, the economic threshold is much higher than the density effect threshold, then control is unnecessary because the population is self-limiting at densities below the economic threshold. At intermediate levels, it is necessary to incorporate the non-linear feedback loop of Figure 6 and run the model on a computer.

CONCLUSION

I have proposed that pest problems can be grouped into paradigms according to the descriptive mathematical techniques employed. Transfer functions may be appropriate for modelling one such paradigm. However, the population process described was free from interactions with other species, whether plant, parasitoid, predator or competitor, and unaffected by temperature variations. Further work is required to incorporate all of these interactions in the model structure, while retaining the conceptual simplicity of the approach.

In cases where linear elements can mimic the major features of population change, such as the case described, it is possible to determine analytical solutions to the pesticide timing problem. Two solutions have been presented. The economic threshold solution determines pesticide application on the basis of pest recruitment monitoring. The optimal control solution determines pesticide application on the basis of prior knowledge regarding pest recruitment. When non-linear elements are added to the model it may not be possible to determine analytical solutions of this form. For this reason, it is important to discover which non-linear elements are responsible for major qualitative changes in the conclusions and which are not.

Acknowledgment

I thank my colleagues in the Environmental Management Unit, Silwood Park, without whose perceptive criticisms this work could not have progressed. I also thank A.B.S. King for his permission to use unpublished data.

REFERENCES

Auslander, D.M., G.F. Oster, and C.B. Huffaker (1974), *Dynamics of Interacting Populations*, J. Franklin Inst. 297, 5, 345-376.

Bellman, R. (1957), *Dynamic Programming*, Princeton University Press, New Jersey.

Birley, M.H. (1977a), An estimation, simulation and control theory for pest populations in a seasonal environment, Ph.D. Thesis (submitted).

Birley, M.H. (1977b), The estimation of insect density and instar survivorship functions from insect census data (unpubl).

Conway, G.R., G.A. Norton, N.J. Small, and A.B.S. King (1975), A systems approach to the control of the sugar cane froghopper, in *Study of Agricultural Systems*, G.E. Dalton, ed., Applied Science Publishers.

Fewkes, D.W. (1969), The control of froghoppers in sugar cane plantations, in *Pests of Sugar Cane*, J.R. Williams, J.R. Metcalfe, R.W. Mungomery, and R. Mathes, eds., Elsevier, Amsterdam, 309-324.

Hacker, C.S., D.W. Scott, and J.R. Thompson (1975), A transfer function forecasting model for mosquito populations, *Can. Ent.*, 107, 243-249.

Hubbell, S.P. (1971), Of sowbugs and systems: the ecological bioenergetics of a terrestrial isopod, in *Systems Analysis and Simulation in Ecology*, B.C. Patten, ed., 1, Academic Press, 269-321.

Kowal, N.E. (1971), A rationale for modelling dynamic ecological systems, in *Systems Analysis and Simulation in Ecology*, B.C. Patten, ed., Academic Press, 1, 123-192.

Lefkovitch, L.P. (1965), The study of population growth in organisms grouped by stage, *Biometrics*, 21, 1-18.

Leslie, P.H. (1945), On the use of matrices in certain population mathematics, *Biometrika*, 33, 183-212.

Leslie, P.H. (1948), Some further notes on the use of matrices in population mathematics, *Biometrika*, 35, 213-245.

Lewis, E.G. (1942), On the generation and growth of a population, *Sankhya: The Indian J. of Statistics*, 6, 93-96.

Lewis, E.R. (1972), Delay-line models of population growth, *Ecology*, 53, 797-807.

Lewis, E.R. (1976), Applications of discrete and continuous network theory to linear population models, *Ecology*, 57, 33-47.

Lotka, A.J. (1939), *Theorie Analytique des Associations Biologigues, Part II: Analyse demographique avec application particuliere a l'espece humaine*, Hermann, Paris.

Manly, B.F.J. (1974), Estimation of stage specific survival rates and other parameters for insect populations developing through several stages, *Oecologia*, 15, 277-285.

Milsum, J.H. (1966), *Biological Control Systems Analysis*, McGraw-Hill.

Norton, G.A. (1976), Analysis of decision making in crop protection, *Agro-Ecosystems*, 3, 27-44.

Oster, G., and Y. Takahashi (1974), Models for age-specific interactions in a periodic environment, *Ecological Monographs*, 44, 483-501.

Rhodes, E.C. (1940), Population mathematics 1,2, and 3, *Jl. R. Statist. Soc.*, 103, 61-89, 218-245, 362-387.

Sinko, J.W., and W. Streifer (1967), A new model for age-size structure of a population, *Ecology*, 48, 910-918.

Southwood, T.R.E., and G.A. Norton (1973), Economic aspects of pest management strategies and decisions in *Insects: Studies in Population Management*, P.W. Geier, ed., Ecol. Soc. Aust., Memoirs 1, Canberra, 168-184.

Stern, V.M., R.F. Smith, R. Van den Bosch, and K.S. Hagen (1959), The integrated control concept, *Hilgardia*, 29, 81-101.

Usher, M.B. (1972), Developments in the Leslie Matrix model, in *Mathematical Models in Ecology*, J.N.R. Jeffers, ed., Symp. Brit. Ecol. Soc., Blackwells, Oxford, 29-60.

Vandermeer, J.H. (1975), On the construction of the population projection matrix for a population in unequal stages, *Biometrics*, 31, 239-242.

THE CONTROL OF ADAPTABLE PESTS

H. N. Comins

INTRODUCTION

An ecological approach to the problem of pest control requires that management decision rules should be based on a knowledge of the population dynamics of the crop-pest system. In addition, the ideal decision criterion should be easy to apply, require little data collection and ensure consistently good results. Reconciling these objectives calls for a great deal of ingenuity, since ecological systems rejoice in a wealth of complexity.

From the pest management point of view the most important single characteristic of biological populations is their ability to adjust to any change in their environment. This is a useful ability in the crop plant, enabling it to recover from early pest attack with little loss of yield. The same ability in the pest, however, results in a continual battle to retain effective control.

This problem has both ecological and evolutionary components. The adaptations which enable organisms to exploit ephemeral habitats (the so-called r-strategy, Conway, 1976) are particularly troublesome. They include the tendency to migrate (causing reinfestation), and a very large reproductive rate, which allows quick recovery from control measures. This recovery is often exaggerated by the deleterious effect of pesticides on insect predators (De Bach, 1974).

The most obvious evolutionary response to control measures is the development of pesticide resistance, although similar responses to other techniques (e.g. sterile male release) are quite conceivable. There are only two ways to avoid this problem: to exterminate the pest species or to set it a synthetic problem beyond possibility of solution. The former is a practical alternative in very few cases, while the latter is made extremely difficult by the requirement that a pesticide should be harmless to other species. It seems, therefore, that pesticide resistance must generally be regarded as inevitable and all the skillful manager can attempt to do is extend the useful lifetime of each pesticide.

THE MANAGEMENT OF PESTICIDE RESISTANCE

The development of resistance in a pest population proceeds via the selective removal of susceptible individuals during pesticide treatments. Thus, in addition to the obvious effect of a (temporary) reduction in the pest population, each treatment results in an increment in the frequency of resistance genes. In the case of mono-factorial resistance we can define the "susceptibility" of the population to be $\ln(p)$, where p is the frequency of the resistance gene (Comins, 1977a). Then, in applying a given strength pesticide treatment, we not only use up a certain amount of money, but also a certain amount of susceptibility (see Figure 1). The latter may be regarded as a free but irreplaceable natural resource; once it is exhausted the pests will be resistant to the given pesticide, and a more expensive substitute must be found.

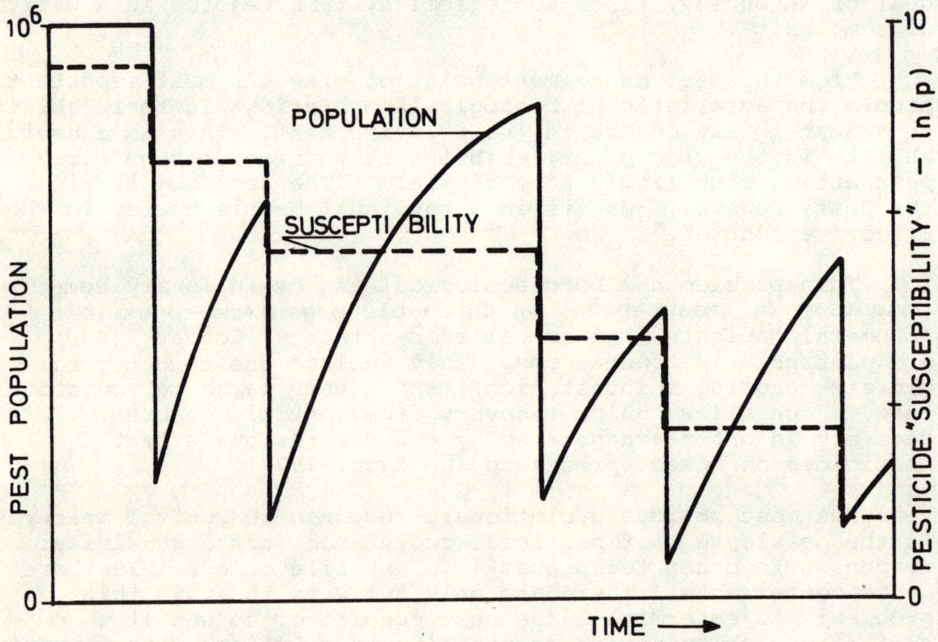

Figure 1. Illustration of the changes in pest population and pesticide susceptibility ($-\ln(p)$, p=frequency of resistance gene) produced by a series of pesticide treatments.

We might therefore ask the following question. Suppose we have a situation in which the control strategy has been optimized to deal with the ecological responses of the pest. How should this strategy be modified to take account of the necessity for conserving susceptibility? Clearly this would not involve giving up the pesticide completely, since susceptibility to an unused pesticide is worthless. However, it might require a more sparing usage, or even toleration of less severe pest attack.

For this purpose I elected to study the class of problems typified by the sugar cane froghopper system, which has been thoroughly investigated by our group (Conway, Norton, Small and King, 1975). The froghopper has four generations per year, of which the first and last are relatively insignificant. As a first approximation to the population dynamics, we assume a standard density-dependent relationship between the sizes of consecutive broods:

$$\frac{n_{t+1}}{N_{t+1}} = \left(\frac{n_t}{N_t}\right)^{c_{t+1}} \quad (1)$$

where n_t, n_{t+1} are the actual brood sizes

N_t, N_{t+1} are "equilibrium" brood sizes

c_{t+1} is the density-dependence coefficient .

The coefficient $\underline{c}$ measures the degree of linkage between the brood sizes; for $c=0$ the second brood size is independent of the first. The usual range is $-1 < c < 1$, where $0 < c < 1$ represents undercompensating density-dependence (imperfect recovery from a disturbance to the previous brood) and $-1 < c < 0$ represents overcompensation, such as might occur if an insect predator were adversely affected by the pesticide.

We measure the size of the brood in terms of adult-days (i.e. the area under the peak on a population versus time graph). This is also a reasonable first estimate of the damage-causing ability of a brood, and we therefore assume that the monetary loss caused by a particular brood is proportional to $\underline{n}$; that is,

$$D_t = e_t \, n_t \quad (2)$$

where e_t = \$ loss per adult-day.

Given the equilibrium populations for a given sized plot and the corresponding cost of pesticide treatment, we can now derive the optimal control strategies for a given assumption about the $\underline{c}$'s. Some examples are shown in Figure 2.

In the interests of tractability, a number of assumptions and approximations are necessary when resistance effects are incorporated into this model. These are summarized in Table 1.

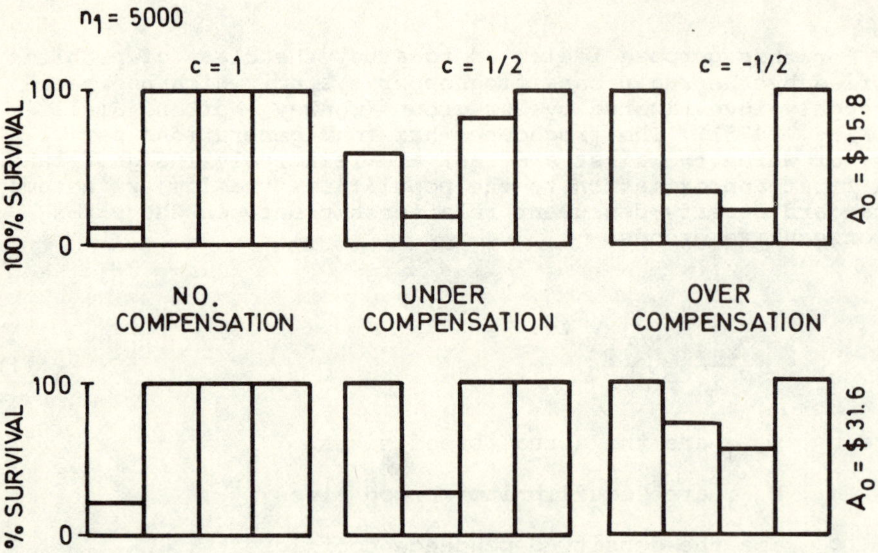

Figure 2. Optimal fixed control strategies for the froghopper with first brood size n_1=5000, density dependence coefficients $c_2=c_3=c_4=c$, and pesticide application cost $A=-A_o \ell n (s)$ (s = fractional susceptible survival, A_o in Trinidad and Tobago dollars, see Equation 3). The strategies are expressed in terms of percent susceptible survival for the four broods, this being a measure of the intensity of pesticide treatment.

Table 1. Assumptions & approximations required for including resistance in the froghopper control model.

GENETICS	1. Heterozygote selection.
	2. Negligible migration.
	3. Single non-recessive gene. Polyfactorial case should be essentially similar provided recessive effects absent.
	4. As treatment strength is varied heterozygote survival is related to susceptible survival by a power law: $s_h = s^{1-\beta}$, $0<\beta<1$.
ECONOMICS	5. The cost of resistance is a decreasing function of the amount of time elapsing before it becomes endemic.
	6. The final phase of selection, where resistance is detectable (so that the frequency could be used to guide management) is of comparatively short duration and need not be considered.
	7. The cost of a pesticide treatment is proportional to the log susceptible survival. This implies that cost is proportional to kill for small kills, but increases steeply for large kills.

The main genetic assumption is that resistance is spread by heterozygote selection, a much faster process than the homozygote selection required for recessive genes (similar slow selection occurs when a group of genes confers resistance, but each separate gene has no effect). In addition we assume negligible migration, since this can greatly delay the selection of resistance if it exceeds a critical value (Comins, 1976).

An analytic treatment of the resulting model (Comins, 1977a) yields the very simple result that the best strategy for maximizing profits, including costs resulting from resistance, is given by regarding the pesticide as effectively more expensive, but otherwise ignoring the phenomenon of resistance. Thus, if an optimization method has been constructed for the no-resistance model, it can immediately provide optimal strategies in the presence of resistance.

The simplicity of this result is a consequence of the fact that it can be derived solely from the genetic and economic assumptions, without regard to the population dynamics. Note that as it stands it does not give the actual value of the effective pesticide price. Thus we can derive only the general form of the optimal strategy rather than the specific result.

The appropriate value for the effective pesticide price is that which minimizes expected total cost. To find this we need to make specific assumptions regarding the resistance cost function and the initial frequency and rate of selection of the resistance gene. The correct price can then be found graphically (see Figure 3). In practice only a crude result can be obtained, since the required data can only be estimated.

FORMULATION OF DECISION RULES

The results of the resistance model have so far been expressed in terms of a fixed schedule, control strategy. In fact a "spin-off" of the resistance work was a method of calculating these strategies analytically (Comins, 1977a).

An alternative method of formulating the results of the model would be as a set of economic thresholds. This does not require any modification of the actual model, but might be expected to provide more consistently good results from management, since the economic threshold procedure uses feedback from field data to correct the model's population predictions.

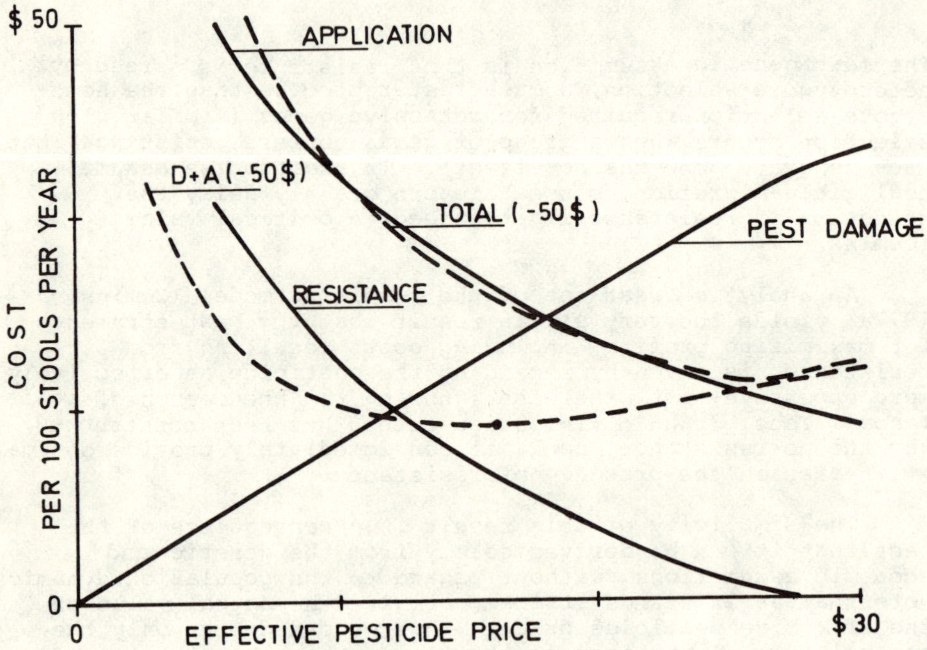

Figure 3. Graphical determination of the optimal effective pesticide price using a resistance cost function with a ten-year time horizon (from Comins 1977a). Prices are in Trinidad and Tobago dollars.

For systems of this general type, the economic threshold for a single monitored variable can be derived by a simple graphical technique, allied to dynamic programming (Comins, 1977b). There are also correspondences between the economic thresholds and fixed schedule optimal strategies which enable one result to be converted to the other; converting economic thresholds to fixed schedules is, however, far easier than the reverse process.

The resistance results actually require a rather novel form of economic threshold, since we assumed an infinity of possible control actions whose cost depended on their effectiveness:

$$A = A_o(-\ell n\ s) \tag{3}$$

where A = treatment cost,

A_o = pesticide price for 37% ($=e^{-1}$) survival,

s = fractional susceptible survival.

Since a continuum of control actions is available, the optimal strategy changes continuously with the size of the pest population. The economic thresholds (as shown in Figure 4) are the populations above which control actions of more than a given severity are worthwhile. Also a control action of this particular severity is optimal when the population is exactly equal to the threshold.

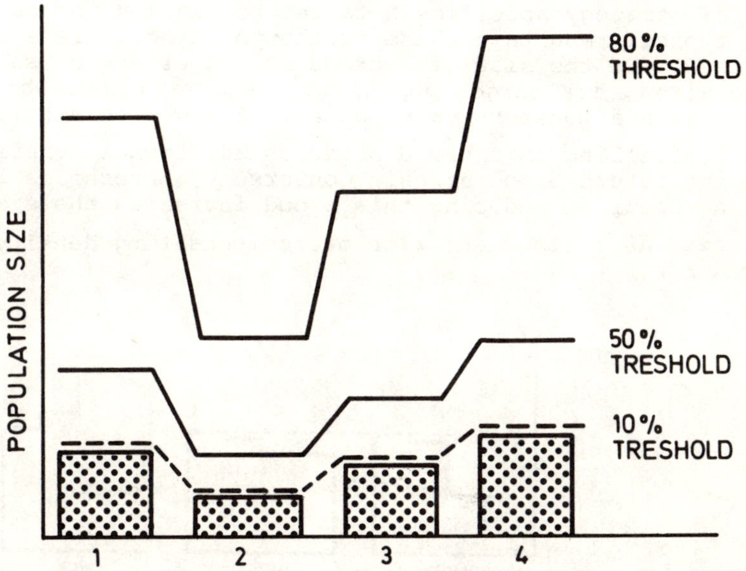

Figure 4. Illustrates the simple relation between economic thresholds for different strength treatments which enables the separate threshold criteria to be replaced by a single "target population". This is a consequence of the logarithmic cost-effectiveness relationship (Eq. 3).

For the cost function (Equation (3)) the thresholds are very simply related—the threshold for removing all but a fraction <u>s</u> of the pests is

$$E(s) = N_o/s \tag{4}$$

where N_o is a constant. Since leaving a fraction s is the optimum strategy when the population is E(s), it follows that by using the optimum strategy we are invariably left with a residual population of size N_o (illustrated in Figure 4 as dotted areas). Thus, instead of specifying the separate economic thresholds, we could condense them into a single decision rule, which requires that whenever the pest population exceeds a target value, one should use the control action which reduces it to the target.

The economic thresholds vary from brood to brood, so the control strategy specifies a target population for each brood. Some typical results for the froghopper system are shown in Figure 5. If the size of a brood has no effect on subsequent brood sizes, its target population is always that which does damage with a monetary value of A_o. A lower threshold means that controlling this brood produces additional benefits in reducing future brood sizes. Conversely, thresholds larger than A_o occur if reducing this brood increases the size of the next, as is the case with overcompensating density dependence.

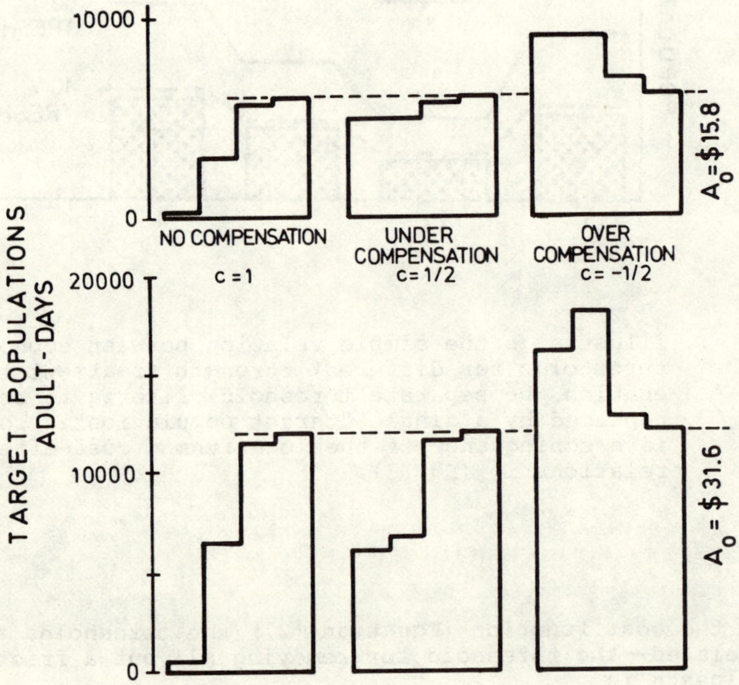

Figure 5. Formulation of the results of Fig. 3 as "target populations". These results apply to any first brood size.

In practice this strategy should be modified to exclude very low strength pesticide treatments. Equation 3 does not include the fixed costs which make such weak treatments uneconomic.

CONCLUSION

The resistance work reported here is based on the philosophy that it is as good to provide an exact answer to an approximate problem as to provide an approximate answer to the exact problem. It is hoped that the results will provide a framework for investigating the effects on pest management of changes in the genetic, population dynamic and economic assumptions.

REFERENCES

Comins, H.N. (1976), The development of insecticide resistance in the presence of migration, *J. Theor. Biol.*, in press.

Comins, H.N. (1977a), The management of pesticide resistance, *J. Theor. Biol.*, in press.

Comins, H.N. (1977b), Economic thresholds for optimal timing of pest control, in preparation.

Conway, G.R., G.A. Norton, N.J. Small, and A.B.S. King (1975), A systems approach to the control of the sugar cane froghopper, in *Study of Agricultural Systems*, G.E. Dalton, ed., Applied Science Publishers, London, 193-229.

Conway, G.R. (1976), Man Versus Pests, in *Theoretical Ecology*, R.M. May, ed., Blackwell, Oxford, 257-281.

De Bach, P. (1974), *Biological Control by Natural Enemies*, Cambridge University Press, London.

KOCHI PREFECTURAL INSTITUTE OF AGRICULTURAL AND FOREST SCIENCE, DIVISION OF ENTOMOLOGY INO, AGAWA-GUN, KOCHI 781-21, JAPAN

INTRODUCTION:

Project team for integrated control of rice pests:

Leader: K. Kiritani

Staff: Y. Kawahara
T. Sasaba
F. Nakasuji

This project team was organized in 1966 to develop an integrated control program for rice insect pests, with particular emphasis being placed on the green rice leafhopper. The research fund is entirely provided by the Ministry of Agriculture and Forestry. This ensures that the fundamental research carried out is relatively independent of the policy of the local government.

Our group is entirely composed of population ecology-minded entomologists. This group structure has determined, rightly or wrongly, our type of systems approach. However, we are confident that fundamental research is the shortest way to solve practical problems and that the systems-oriented approach, where possible, is the best means of designing pest management programs where integrated use of various measures is necessary.

A SYSTEMS APPROACH TO PEST MANAGEMENT OF THE GREEN RICE LEAFHOPPER

K. Kiritani

INTRODUCTION

The panel convened in 1975 to discuss the status of rice pests in Japan ranked the green rice leafhopper (hereafter GRL), *Nephotettix cincticeps* UHLER, as the number one key pest in southern Japan. The damage caused by GRL is due to transmission of Rice Dwarf Virus (RDV) to rice plants. Although direct damage is inflicted on rice through its sucking behavior, it is considered to be negligible, at least in southern Japan (Nakasuji and Nomura, 1968; Kassai and Ozaki, 1972).

In southern Japan, GRL has five generations, and overwinters as mature nymphs. RDV disease is transmitted to rice plants by infective nymphs and adults. Since it is a persistent type of cirus, individuals of GRL become infective through transovarial transmission and/or feeding acquisition of RDV from infected rice plants. During the pre-war period, outbreaks of RDV occurred only sporadically in southern Japan, lasting at the most for one to two years. Since around 1955, however, this disease has become more prevalent, the result of an increase in the population density of GRL, and it has inflicted considerable losses on rice yield (Nakasuji and Kiritani, 1976).

The RDV epidemic is considered to have resulted from the following factors (Nakasuji and Kiritani, 1976): (1) Increased cultivation of early sown rice, providing a favorable host for the first generation which would otherwise develop on gramineaceous weeds in the fallow paddy field. (2) Increase in the acreage of fallow paddy fields, resulting from a decrease in the cultivation of such winter crops as wheat and barley. This provides overwintering nymphs of GRL with a large quantity of winter hosts. (3) Decimation of natural enemies of GRL by the use of such broad spectrum insecticide as BHC, parathion etc., to control the rice stem borer, *Chilo suppressalis*. Specifically, BHC was so toxic to spiders that even the granular formulation of BHC affected wolf spiders, *Lycosa pseudoannulata* BOES. ET STR., through food-chain toxicity (Kiritani and Kawahara, 1973). (4) The development of a high level of insecticide resistance to various kinds of organophosphates and carbamates. These conditions make GRL the principal target of pest management systems of rice pests in Japan.

LIFE HISTORY OF GRL

In the early paddy area, rice is grown from the middle of March, a second crop of rice being grown in a limited area in Kochi after the harvest of the first crop. In the mid-season paddy area, rice is cropped once a year, being sown in early May and harvested in late September.

Overwintering nymphs, feeding on graminaceous grasses in the fallow, reach the fifth instar in February and emerge as adults in April. In the early paddy area, they emigrate to paddy nurseries for egg laying. Here, GRL completes two consecutive generations on each crop, further development of the population being interrupted by harvest. In the mid-season paddy area, a further generation, following the overwintering generation, occurs on grasses, such as foxtail, adults of this generation invading paddy fields transplanted in mid-June. After two consecutive generations on the mid-season rice, adults of the fourth generation oviposit their eggs on rogue rice plants and grasses in the fallow, most of the hatchlings (fifth generation) developing to third instar nymphs before winter. The peaks of adult emergence in the first, second, third, and fourth generations are observed in mid-June, late July, late August, and late September, respectively. The emergence of fifth generation adults (the overwintering generation) occurs in the following spring, from mid-March to mid-April.

A SYSTEMS APPROACH TO GRL MANAGEMENT

The management of GRL populations can be regarded as a long-range plan, and it is felt that the construction of a simulation model is essential in developing a management system for this species. Consequently, our study, which was initiated as a life-table approach to GRL (Kiritani et al. 1970), has developed into a life system approach, where natural enemies and RDV are dealt with as independent components in the environment, not as mere environmental constraints to GRL. As a result, an RDV epidemiology model (Nakasuji and Kiritani, 1972; Nakasuji et al., 1975), a life-system model of GRL involving spiders (Sasaba et al. 1973; Sasaba and Kiritani, 1975), a predator-prey system model (Sasaba, 1974; Kiritani and Kakiya, 1975; Kiritani, 1977; Miyai et al. in press) and population dynamics models of spiders (Kawahara et al., 1974; Kawahara, 1975) have been developed.

THE RDV EPIDEMIOLOGY MODEL

Structure of the Model

Nakasuji and Kiritani (1972) have described a systems model for the natural spread of RDV-infestation. The model is composed of sub-models that describe changes in the population density of GRL during the breeding season and changes in the percentage of infected insects among the vector population. Sub-models also describe the process of transmission of RDV to rice plants by the vector and the feeding acquisition of RDV by the vector from infected rice plants (Figure 1). The relationships between these sub-models and the essential parts of the calculation procedure are illustrated for the early sown paddy in Figure 2.

As described in the following section, Sasaba and Kiritani (1975) have developed a systems model that describes the population dynamics of GRL, including its overwintering period. Hence, a new epidemiological model of RDV was developed by substituting Sasaba and Kiritani's model for the population model in the aforementioned RDV epidemiology model (Nakasuji et al. 1975). The new model not only permits a continuous calculation to be made over successive years; but also allows evaluation of the role that individual factors play in determining the prevalence of RDV.

Simulation

Changes in the percentage of infected rice plants and loss in yield of husked rice were examined by changing the population density of GRL and the percentage of infected insects (Nakasuji and Kiritani, 1972) (Table 1). It was concluded that the population density of the vector in the overwintering generation must be decreased to less than one-tenth, or the transmission of RDV by the vector of that generation to the nursery of early sown rice should be decreased to less than one-third, if the percentage loss in yield is to be reduced to less than 10 percent.

In addition, simulation tests, using the new model, were conducted to examine the effect of the following factors on the prevalence of RDV: vector density, the efficiency of feeding acquisition of RDV by the vector, and the efficiency of RDV transmission by the vector to the rice plant (Nakasuji et al., 1975). This was done by determining the effect on infected insects and loss in yield obtained by reducing each value of the parameters to one-half of the value in Kochi Prefecture (Table 2). This simulation indicated that the most influential parameter is the efficiency of transmission. Regrettably, all of the *japonica* varieties of rice are believed to be susceptible to GRL and RDV (Ishii et al. 1969).

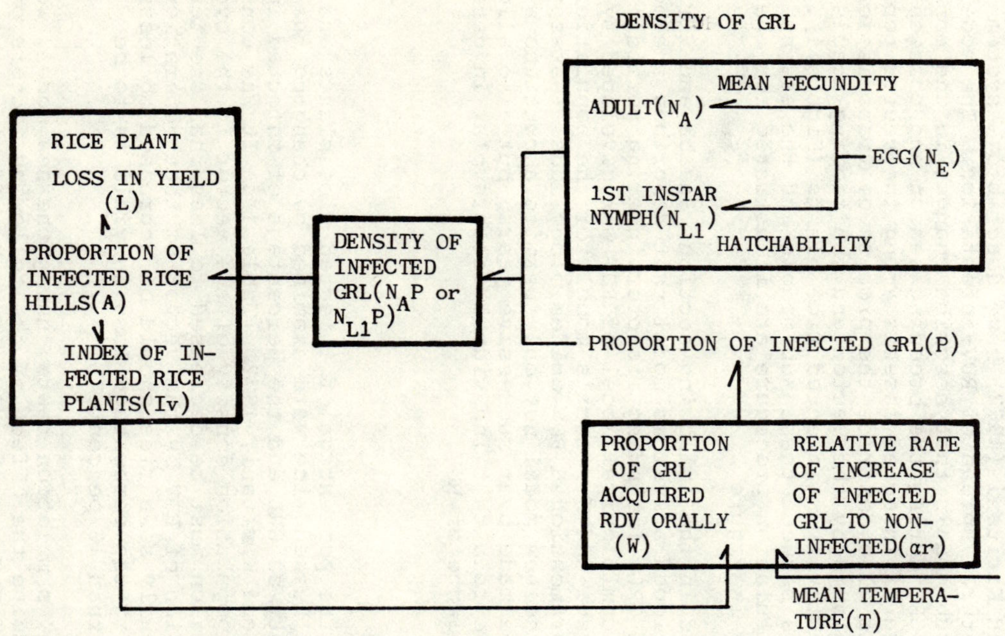

Figure 1. Schematic representation the structure of RDV epidemiology model.

Figure 2. Major components of RDV model.

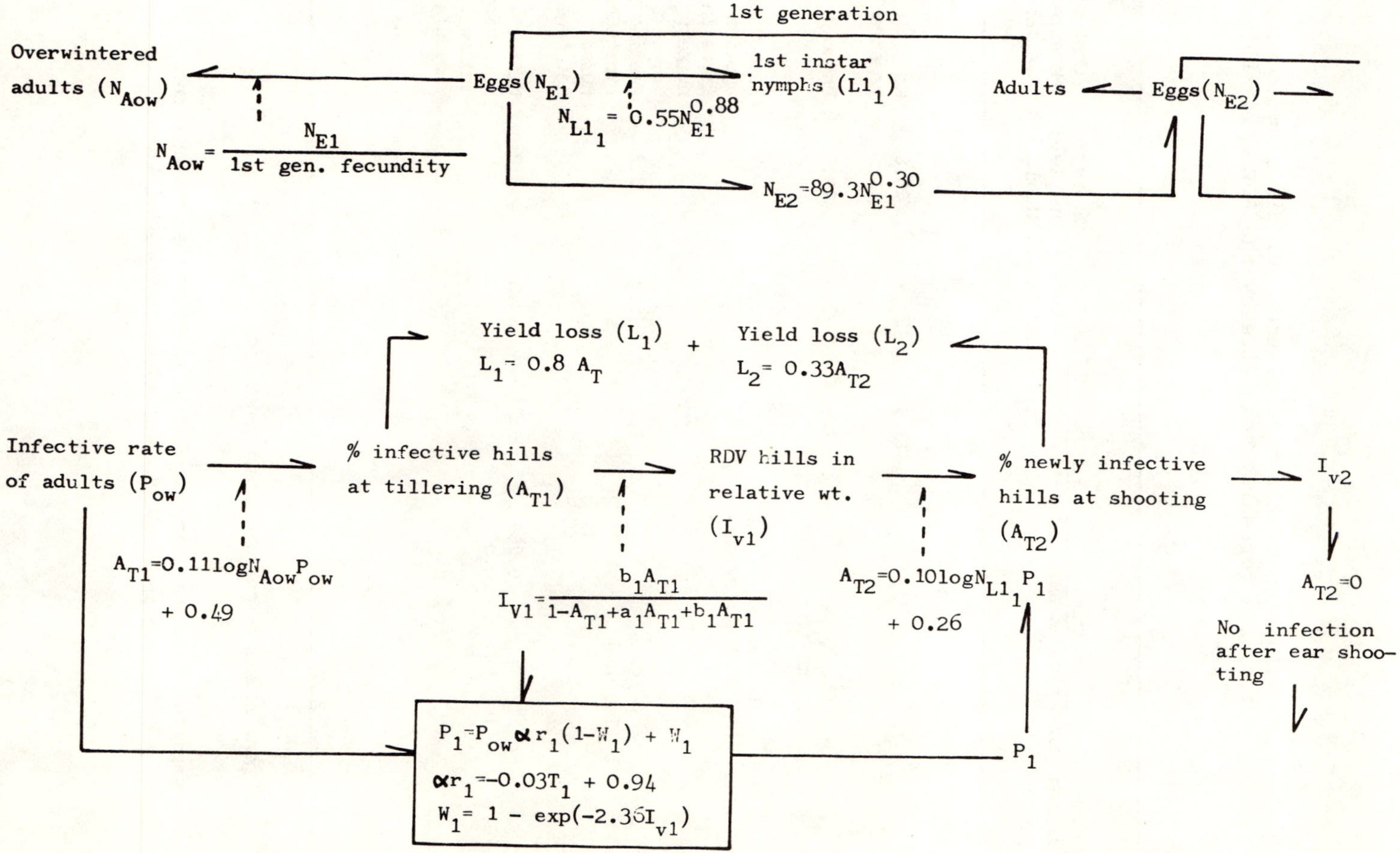

P: rate of infective insects, d: relative survival **rate** of infected insects,

r: rate of transovarial transmission, T: mean temperature,

W: rate of oral acquisition of RDV

Table 1. Evaluation of the effectiveness of different control measures for early sown rice by the RDV epidemiology model.

Simulation	Treatment: infective	Treatment: density of	% infected hills at tillering (A_I)	% infected hills at shooting (A_{II})	% loss in yield (L_I+L_{II})	Control Measures Available
1	1	1	13.3	29.9	16.8	No treatment
2	1/3 (ow.gen) [1)]	1	7.8	22.6	11.6	Unknown
3	1	1/3 (ow.gen)	7.8	19.6	10.4	Insecticide applications in nursery
4	1/3 (1st gen)	1	3.2	16.9	7.2	Feeding inhibition in nursery with net cage
5	1/3 (1st gen)	0.08 (ow.gen) [2)]	3.2	4.3	3.0	Winter ploughing of fallow fields; aerial broadcasting of insecticide

[1)] ow. gen. refers to overwintering generation

[2)] Decimation of overwintering population to 8% of the untreated one will result in 2/3 decrease of the P in the first generation

Table 2. Effects of parameters in the RDV epidemiology model on the percent of infective insects and yield loss when the value of each parameter is reduced to one-half.

Parameter	Infective Insects (%)	Yield Loss (%)	Suggested Control Measures
Density of GRL	6.2	18.1	Insecticide treatment or winter ploughing of fallow paddy fields
Efficiency of feeding acquisition of RDV	3.0	16.9	Genetic manipulation of GRL
Efficiency of transmission of RDV	3.0	8.5	Utilization of varietal resistance

Forecasting of RDV Infestation

From a practical point of view, the main disadvantage of these RDV models is that the calculation starts after the completion of egg laying in rice fields by immigrated adult females. In practice, it is important to know the percentage of RDV infected hills before the occurrence of adult immigration.

In areas of middle-season rice cultivation, the density of GRL in the first generation can be monitored by net sweeping fallow paddy fields in late May. The number of adults per hill which invaded paddy fields (N_{A1}) is given by multiplying the number of insects obtained by 10 double sweepings in fallow paddy fields (N_{Fal}) by the coefficient 0.005 (Horikiri et al., unpublished). The percentage of infected insects in the overwintering generation (P_{Ow}) can be measured until late May either by the serological method or by feeding tests. The percentage of infected insects of the first generation (P_1) decreased to 60 percent of that in the overwintering generation in middle-season rice, i.e. $P_1 = 0.6\ P_{OW}$ (Nakasuji and Kiritani, 1971). Hence, the percentage of infected rice hills at the tillering stage (A_T) can be estimated by the following equation:

$$A_T = 0.11\ \log N_{A1} P_1 + 0.49$$ (Nakasuji and Kiritani, 1972)

Control Threshold of RDV

Assuming random invasion of infective adults to the rice field, the control threshold (N), expressed in terms of the population density of GRL per hill, can be determined by

$$A_T = 1 - \exp(-aNLP) \tag{1}$$

where A_T is the proportion of infected hills, a is the number of hills infected by an infective adult per day, N is the number of adults per hill that immigrated into the rice field, L is the mean longevity of adults under natural conditions and P is the proportion of RDV infective individuals (Kiritani and Nakasuji, unpublished). Fifteen percent RDV infective hills, including infection at tillering and shooting stages, is considered to be the level where some yield loss will occur (Sugino, 1975).

By substituting the following values tentatively in Eq. 1, 0.15 for A_T (Sugino, 1975), 1 for a (Ishii, personal communication), 10 for L (Hokyo, personal communication), and 0.05 for P, which is an average value for the first generation in Kochi (Nakasuji, 1974), we get N=0.32 adults/hill as the control threshold density of GRL.

Expressed in terms of the number of insects per hill per day over the invasion period of 30 days, N can be expressed as 0.32 x 10/30 = 0.1 adults/hill/day. Such a low density as one adult per 10 hills (when P=0.05) imposes difficulty in controlling RDV infection, requiring heavy use of insecticide.

THE LIFE-SYSTEM MODEL OF GRL

The life-table study of GRL showed that the population density of GRL is regulated through negative feed back mechanisms that come into operation at different densities, and that escape of the population from one mechanism will automatically bring into action other mortality factors. Regulatory processes that operate at the adult stage, i.e. dispersal and reduction in fecundity, contribute most towards stabilizing the fluctuation of the GRL population. It was also suggested that the effect of the series of such regulatory mechanisms on density stabilization was not always perfect within a generation, and that stabilization was achieved through successive generations, e.g. two generations in GRL (Kiritani et al., 1970).

As mentioned in the introduction, outbreaks of GRL are partly attributable to the decimation of spider populations in paddy fields by the use of insecticides which include BHC.

Therefore, predation of GRL by spiders would appear to be an essential component of the life-system of GRL. Of these spiders, Lycosa pseudoannulata is the most voracious predator of GRL in the paddy fields of southern Japan. The predation by this spider on GRL made up to about 90 percent of the total cases of predation in the paddy field. Planthoppers usually appear in the diet of Lycosa late in the season but to a lesser extent compared with GRL (Kiritani et al., 1972; Kiritani and Kakiya, 1975). The "sight-count" method proposed by Kiritani et al. (1972) enabled the parameters of the functional and numerical responses of spiders to GRL density to be determined.

Structure of the Model

A descriptive model of the GRL life-system including predation by spiders, L. pseudoannulata and Oedothorax insecticeps, was constructed on the basis of GRL life-tables (Sasaba et al., 1973; Sasaba and Kiritani, 1975) (Figure 3). The numerical relationship and/or mortality factors associated with the successive developmental stages of GRL are summarized in Table 3 with other parameters involved. The type of population growth and functional response of L. pseudoannulata used in the systems model are represented in Figures 4 and 5, respectively. The number of Lycosa/day/hill(Sd) multiplied by the number of GRL attacked by one Lycosa/day/hill was assumed to be the number of GRL eaten by Lycosa/hill/day.

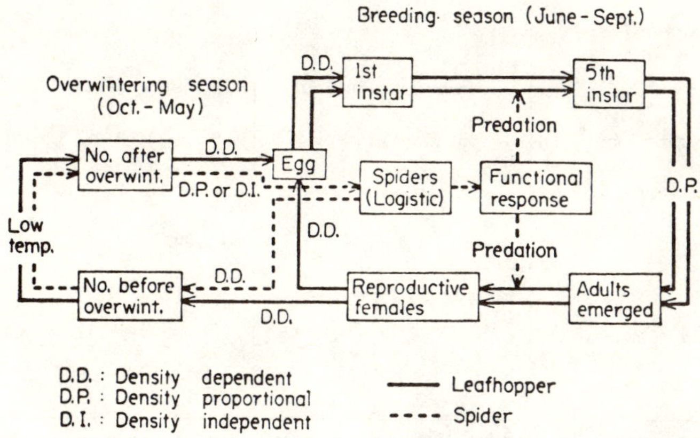

Figure 3. A flow chart of the systems model.

Table 3. Numerical relationships between successive time intervals.

Time interval (Symbol) From	To	Numerical relationship and/or mortality factor	r^2
GRL (Breeding season)			
Egg (E)	1st instar nymph (N_1)	$N_1=0.36E^{0.89}$	0.94
1st instar nymph (N_1)	5th instar nymph (N_5)	Predation* Constant survival rate (S_n)*	
5th instar nymph (N_5)	Adult emerged (A_e)	$A_e=0.77N_5$	0.99
Adult emerged (A_e)	Adult matured (A_m)	Predation* Constant survival rate (S_a)*	
Adult matured (A_m)	Female matured (A)	$A =0.50A_m$	0.99
Female matured (A)	Egg (E)	$E=192.27A^{0.41}$	0.56
GRL (Overwintering season)			
Adult emerged in the 3rd generation (A_3)	Nymph before wintering (N_b)	$N_b=91.60-0.52A_3-4.72L$	0.79
Nymph before wintering (N_b)	Nymph after wintering (N_a)	$N_a/N_b=0.80-0.0045N_b-0.0076T$	0.94
Nymph after wintering (N_a)	Egg in the 2nd generation (E_2)	$E_2=-139.71 \log N_a+229.32 (N_a>3)$ $E_2=54.23N_a \ (0\leq N_a\leq 3)$	0.63
Lycosa spiders (Breeding season)			
Lycosa invaded paddy field (L_o)	Maximum no. *Lycosa* (L_m)	$L_m=7.15L_o+1.67$ $L_t = \dfrac{L_m}{1 + \left(\dfrac{L_m}{L_o} -1\right)e^{-0.1t}}$	
Lycosid spiders (Overwintering season)			
Maximum no. *Lycosa* (L_m)	*Lycosa* before wintering (S_{12})	$S_{12}=-1.04L_m+10.49$	0.97
Lycosa before wintering (S_{12})	*Lycosa* after wintering (S_2)	$S_2/S_{12}=4.17+0.082S_{12}-0.076T$	0.97
Lycosa after wintering (S_2)	*Lycosa* invaded paddy field (L_0)	$L_0=0.0062S_2-0.001$	0.89
Microphantid spiders	*Oedothorax* invaded paddy field (O_0)	$O_0=0.80$	
	Maximum no. *Oedothorax* (O_m)	$O_m=-3.15 \log Z+2.29$ $O_t= \dfrac{O_m}{1 + \left(\dfrac{O_m}{O_o} -1\right) e^{-0.1t}}$	0.81

T: Number of days below 0^oC during Dec.-Feb.
L_o: Mean number *Lycosa* in July per hill.
L_t: Number of *Lycosa* per hill on t-th day.
O_0: Mean number of *Oedothorax* in July per hill.
O_t: Number of *Oedothorax* per hill on t-th day.
Z: Mean number of *Lycosa* during July-Aug. per hill.
t: Number of days after transplantation minus 30 days. Hence, t=0 means 30 days after transplantation when appreciable invasion of spiders was observed.
r: Correlation coefficient.
*: Sasaba et al. (1973)

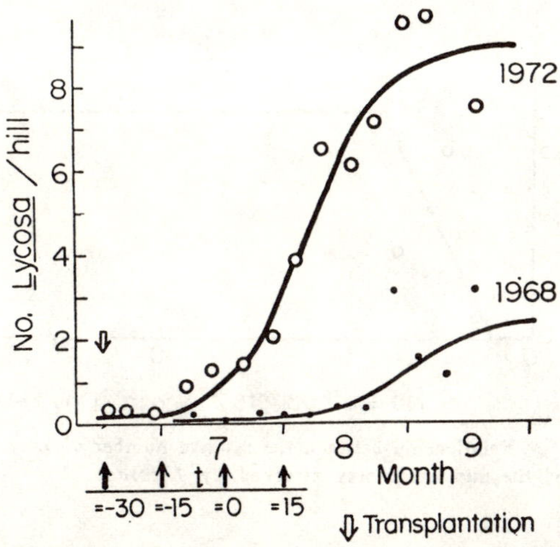

Fig. 4. Sequential changes in number of *Lycosa* per hill. L_m values were 2.5 and 9.0 in 1968 and 1972, respectively. L_o values were 0.05 and 1.0 in 1968 and 1972, respectively. Curves depict the number of *Lycosa* calculated by equation (4) with variables of K and X. Correlation coefficients of observed (ordinate) versus calculated (abscissa) numbers were 0.800 and 0.964 in 1968 and 1972, respectively.

$$L_t = \frac{L_m}{1+\left(\frac{L_m}{L_o}-1\right)e^{-0.1t}} \qquad (4)$$

where L_t: the number of *Lycosa* per hill on *t*-th day

L_m: the maximum number of *Lycosa* per hill

L_o: the mean number of *Lycosa* from $t=-15$ to 15

t : the number of days after transplantation minus 30 days. Hence, $t=0$ means 30 days after transplantation when appreciable invasion of spiders was observed.

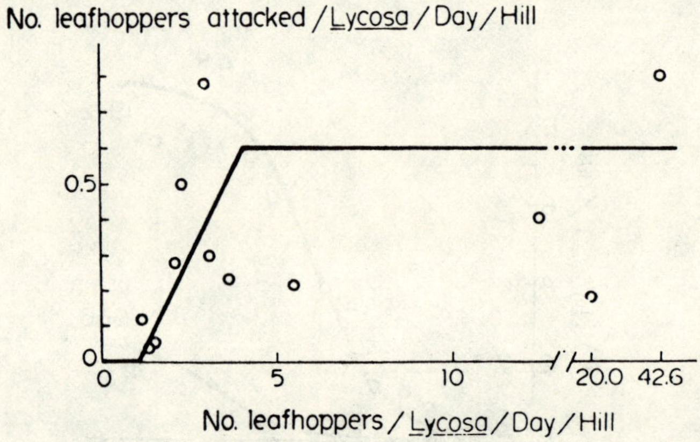

Fig. 5. Relationship between the relative number of prey insects and the number of prey attacked by *Lycosa*.

For ease of calculation, it was assumed that all of the eggs and fifth instar nymphs developed to the next stage simultaneously, and that predation by spiders operated on these cumulative numbers of prey. This calculation was repeated 20 and 7 times, corresponding to the duration of nymphal and pre-ovipositional periods, respectively.

Simulation

The fit of the calculated values to observed ones was examined for the generation-to-generation change in egg density. To obtain initial numbers of GRL eggs (E_2) and Lycosa (L_o), the actual numbers in 1968, i.e. 99.27 and 0.05, were used. Also, the number of days when the minimum temperature recorded was below 0°C was given. As shown in Figure 6, the calculated values coincided with the observed values except for the fourth generation. This deviation in the fourth generation suggests that factors other than those involved in the assumed relationships play an important role in the population regulation of GRL during the breeding season, e.g. differences in the size of hills, climatic factors, etc. These problems remain to be solved in the future.

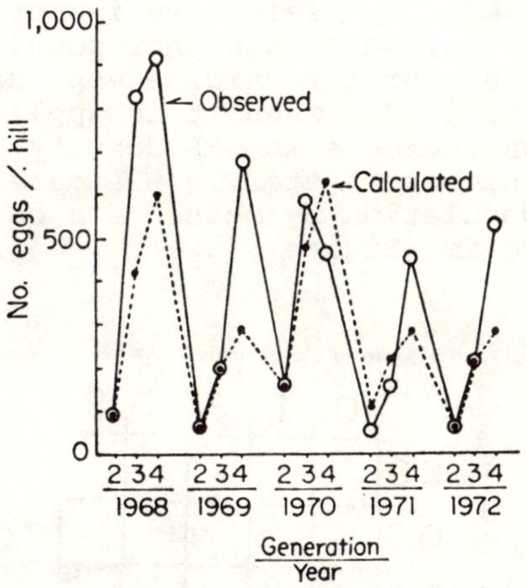

Figure 6. Comparison in the trend of generation-to-generation changes in egg density between observed and calculated values.

Simulation exercises carried out using the GRL life-system model suggested several points of interest from the practical point of view. First, there exists an optimum ratio of predators to prey in utilizing spiders as a biological control agent (Figure 7).

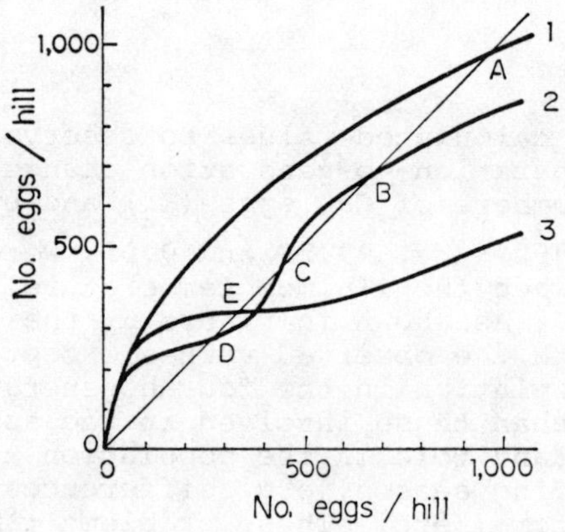

Figure 7. Reproduction curve expressed in terms of the egg density of GRL. As the mean number of *Lycosa* in July per hill (x_1), 0.0, 0.3 and 1.0 were given to curve-1,-2 & -3, respectively.

Secondly, the use of non-selective insecticide, e.g. BHC, induces outbreaks of GRL, while a higher population density of Lycosa reduces the need for insecticide applications. Thirdly, even a selective insecticide, when it is applied at the wrong time, brings about an increase in GRL density. Finally, the model indicates the number of sterile GRL males to be released in suppressing GRL population by means of a combined insecticide and sterile male program (Figure 8).

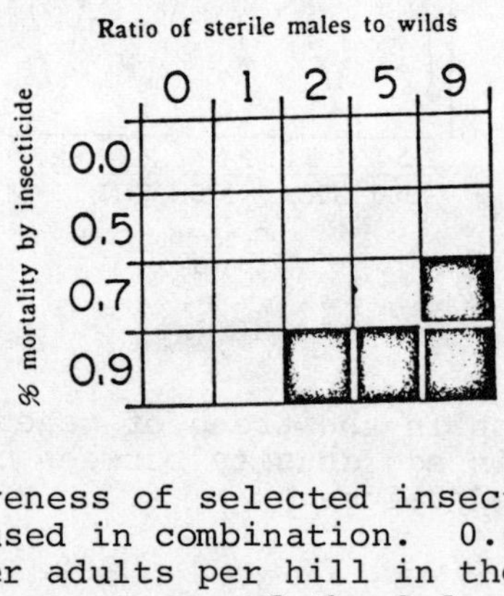

Figure 8. Effectiveness of selected insecticide and sterile-male release when used in combination. 0.40 was given as the number of leafhopper adults per hill in the 1st generation. The number of leafhopper eggs of the 3rd generation is smaller than 150 per hill in the shaded area.

However, this life-system model incorporates the predation by spiders on GRL in such a unilateral way that while GRL is reduced by predation, this does not affect the *Lycosa* population in turn. Consequently, in order to analyze the *Lycosa*-GRL interaction system in the paddy field, effort has been devoted to the construction of a predator-prey system that uses empirical data.

THE LYCOSA-GRL SYSTEM MODEL

Although predators are one of the most effective agents for the biological control of insect pests, the studies of predator-prey interactions have so far been limited to those conducted under laboratory conditions. Information obtained from studies conducted under field conditions is usually too fragmentary to fully understand the interaction between the predator and its prey. In the first place, this results from the technical difficulty involved in simultaneously estimating the population densities of both predator and prey under natural conditions. Secondly, quantitative evaluation of the effect of predation on prey populations requires sophisticated techniques, e.g. precipitin tests (Kiritani and Dempster, 1973). Finally, the polyphagous habit of most predator species and the fact that mature as well as immature individuals feed on the same prey species makes the evaluation much more complicated.

Apart from field problems, the predator-prey system has been one of the central problems in mathematical modelling since Lotka (1925) and Volterra (1926), and many substitute models have since been proposed (see Rosenzweig, 1971). Ironically, both lines of approach, i.e., field and model studies, have taken their own course, without reference to each other, and few attempts have been made to bring them together. As a preliminary step toward this goal, the interrelationships between such parameters as predation rate, survival rate and densities of predators and prey have been investigated under field conditions. In the analysis, the main emphasis was placed on finding an empirical way of modelling the predator-prey system, starting with observed data.

Structure of the Model

The number of prey (X_4) and predators (Y_4) on the (*n*+1)th week can be given by a vector relationship on the X-Y plane as follows (Figure 9):

$$X_4 = X_1 + (X_3 - X_1) - (X_1 - X_2) \tag{2}$$

$$Y_4 = Y_1 + (Y_3 - Y_1) - (Y_1 - Y_2) \tag{3}$$

where X_1 and Y_1 are the number of prey and predators on the nth week, (X_3-X_1) and (Y_3-Y_1) refer to the number of prey and predators recruited during one week, and (X_1-X_2) and (Y_1-Y_2) refer to the number of prey eaten and the number of predators that die during one week, respectively.

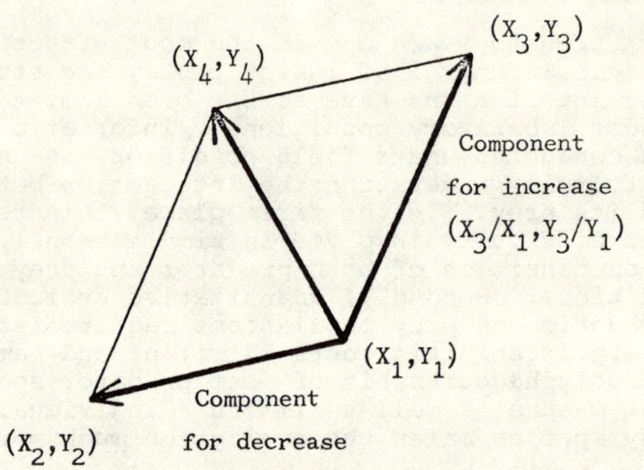

$$\begin{bmatrix}\text{NO.}\\ \text{PREY}\end{bmatrix} = \begin{bmatrix}\text{INITIAL NO.}\\ \text{PREY}\end{bmatrix} + \begin{bmatrix}\text{NO. PREY}\\ \text{RECRUITED}\end{bmatrix} - \begin{bmatrix}\text{NO. PREY}\\ \text{EATEN}\end{bmatrix}$$

or

$$X_4 = X_1 + (X_3 - X_1) - (X_1 - X_2)$$

$$\begin{bmatrix}\text{NO.}\\ \text{PREDATORS}\end{bmatrix} = \begin{bmatrix}\text{INITIAL NO.}\\ \text{PREDATORS}\end{bmatrix} + \begin{bmatrix}\text{NO. PREDATORS}\\ \text{RECRUITED}\end{bmatrix} - \begin{bmatrix}\text{NO. PREDATORS}\\ \text{DEAD}\end{bmatrix}$$

or

$$Y_4 = Y_1 + (Y_3 - Y_1) - (Y_1 - Y_2)$$

Figure 9. Vector relationship of predator-prey system.

The weekly disappearance rate (D) of *Lycosa*, which includes mortality, emigration and, to a lesser extent, loss of individual marks, was obtained from a mark-and-recapture analysis of *Lycosa* adults. The predation by *Lycosa* on GRL was evaluated quantatively by the "sight-count" method (Kiritani et al., 1972). The disappearance rate (D) and the percentage of GRL eaten by *Lycosa* per day (Pr) are expressed by a set of multiple regressions, as functions of the density of GRL (X) and of *Lycosa* (Y) (Kiritani and Kakiya, 1975).

$$\log Pr = -\,0.84\log X + 0.32\log Y + 1.75 \quad (R^2 = 0.73) \qquad (4)$$

$$\log D = -\,0.40\log X + 0.34\log Y - 0.11 \quad (R^2 = 0.98)\ . \qquad (5)$$

Equations (4) and (5) can be transformed into:

$$\text{No. GRL eaten/day/hill} = 0.56X^{0.16}\cdot Y^{0.32} \qquad (4')$$

$$\text{No. disappeared/week/hill} = 0.78X^{-0.40}\cdot Y^{1.34}\ . \qquad (5')$$

Here the powers of Y in Eqs. (4') and (5') denote the degree of interference between lycosid spiders.

The components of recruitment, i.e., (X_3-X_1) and (Y_3-Y_1) in Equations (2) and (3), were measured in terms of X_3/X_1 and Y_3/Y_1, and expressed as functions of X_1 and Y_1 from data pooled for the period 1970-73. The following multiple regressions were derived (Sasaba, 1974),

$$\log(X_3/X_1) = -0.40\log X_1 - 0.03\log Y_1 + 0.73 \quad (R^2=0.59) \qquad (6)$$

$$\log(Y_3/Y_1) = -0.06\log X_1 - 0.10\log Y_1 + 0.42 \quad (R^2=0.23 \qquad (7)$$

or

$$X_3 = 5.37X_1^{0.60}\cdot Y_1^{-0.03} \qquad (6')$$

$$Y_3 = 2.63X_1^{-0.06}\cdot Y_1^{0.90}\ . \qquad (7')$$

Equations (6') and (7') indicate that the values of X_3 and Y_3 are determined mainly by the density of their own species.

From a practical point of view, it is interesting to examine the equilibrium density of GRL without Lycosa. In Equation (6'), the term for Lycosa, $Y^{-0.03}$, takes values nearly equal to one, whatever the value of Y. Hence, by transforming the equation to $X_{n+1} = 5.37\ X_n^{0.6}$, we can obtain the equilibrium density of GRL when Lycosa is presumed absent from the system (Kiritani, 1976). The solution for $X_{n+1} = X_n$ is 67 GRL/m^2 which is 4.3 times that of GRL under the presence of Lycosa. Consequently, the GRL outbreaks during the last 20 years can be explained through the destruction of spider populations by insecticides.

Simulation

It is now possible to simulate the population changes of Lycosa and GRL using Equations (4), (5), (6) and (7). The results of this simulation will be published elsewhere (Kiritani, 1976; Miyai et al., in press). Matching between computer output and empirical data was examined using real data for the initial values of X and Y in running the simulation (Figure 10). In general, the trajectories of calculated values traced relatively well those of empirical data. However, the calculated trajectories failed to mimic variations in observed data along the X axis. As evident from Figure 10, there are two peaks in the GRL density, each corresponding to the second and the third generations. But Equation (6) does not incorporate such a time-dependent component.

The trajectories shown in Figure 11 represent the sequential changes in the two interacting populations at constant time intervals of one week (Sasaba, 1974). The system achieves a stable node, showing direct convergence with no looping from any point in the X-Y plane to the equilibrium point P' ($\overset{*}{X}$ = 15.6, $\overset{*}{Y}$ = 18.4) (Miyai et al., in press). The shaded area represents the domain where densities of Lycosa and/or GRL become negative, in other words, where D>1 and/or Pr>100.

If the number of GRL at the equilibrium point corresponds to a tolerable level of the rice damage, then it would be desirable to have the system brought to that point. Once at the equilibrium point, the system would theoretically stay there, and no further control would be needed, provided that there is no intervention by man, e.g., by harvesting, and no climatic disturbance, e.g., winter. Conventional rice cultivation involves about 20 rice hills per m^2. Given no perturbation from outside, the total incidence of GRL (mature nymphs and adults) would be kept at less than one insect per hill by Lycosa. However, this level is still too high for the practical control threshold of GRL, i.e., 0.1 insect/hill/day when P = 0.05.

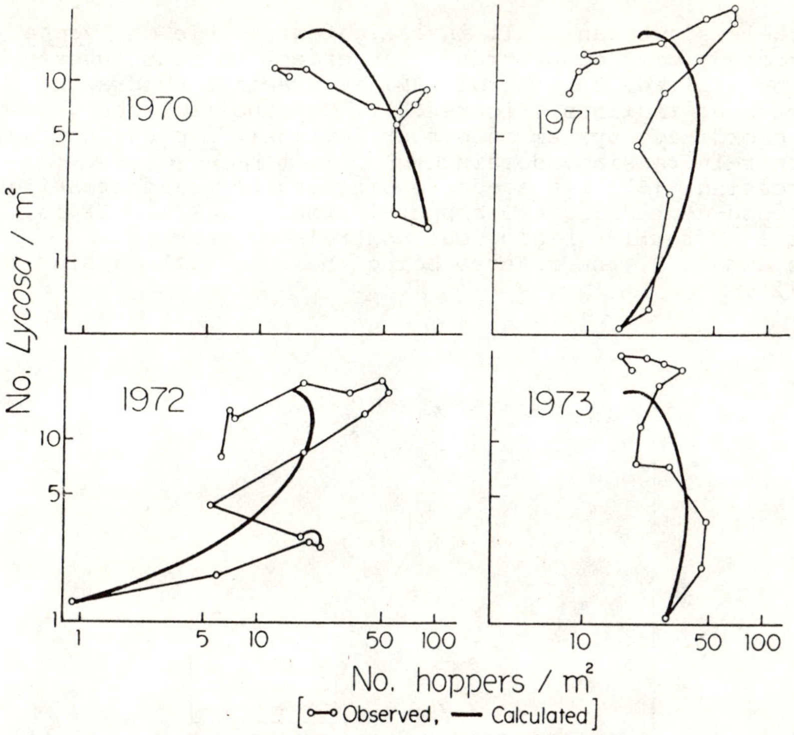

Figure 10. Matching between the computer output and the empirical data using real data for the initial values of X and Y in running a simulation.

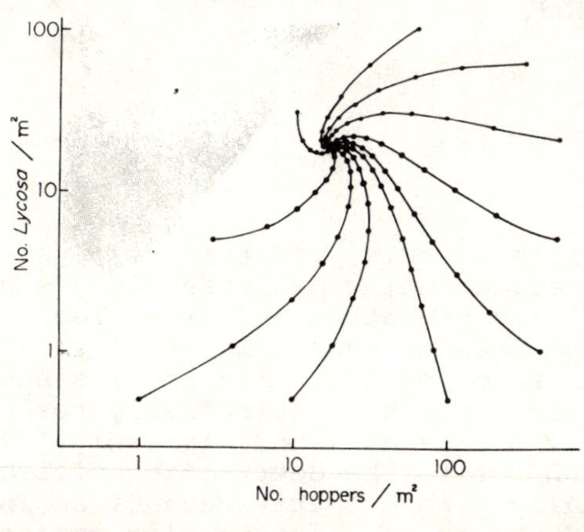

Figure 11. Sequential changes in densities of lycosid spiders and leafhoppers at 7-day intervals predicted by a predator-prey systems model. Shaded area denotes extinction zone of predator and/or prey.

Nevertheless, we can still investigate the role of Lycosa in an integrated, control program. The effective zone, where Lycosa can be utilized to control GRL, is examined by drawing prey and predator isolines (Figure 12). The two isolines divide the coordinate system into four domains: Lycosa-increasing and hopper-increasing, domain (A); Lycosa-increasing and hopper-decreasing, (B); Lycosa-decreasing and hopper-decreasing, (C); and Lycosa-decreasing and hopper-increasing, (D). It is suggested that effective biological control can only be expected in domain B, the reverse being the case with domain D (Figure 12).

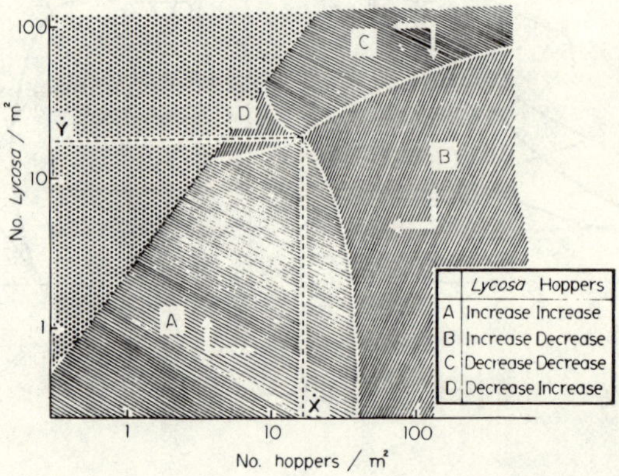

Figure 12. Showing the domain (B) where effective biological control of N.cincticeps by lycosid spiders can be expected.

The susceptibility of this predator-prey system to insecticide was then examined, assuming different levels of insecticide effectiveness in killing predators and prey (Table 4). The simulation shows that insecticide application always shifts the equilibrium point in favor of GRL compared with Lycosa. Simulation of BHC application, which is selectively toxic to Lycosa (Kawahara et al., 1971; Takahashi and Kiritani, 1973), shows the resurgence of GRL due to the destruction of Lycosa (Treatment A in Table 4). Finally, the effectiveness of an integrated program was assessed, using a resistant rice variety as well as spiders in controlling GRL (Table 5). The simulation suggests that a moderate degree of resistance to GRL, about 40 percent suppression of the increment (X_3/X_1 in Equation 6), is sufficient to control GRL.

Table 4. Effect of weekly insecticidal treatments on the *Lycosa* (spider)-*Nephotettix* (leafhopper) system (computer simulation).

Treatment	Effectiveness (% kill)		Equilibrium density (m^2)	
	Lycosa	Nephotettix	Lycosa	Nephotettix
Control	0	0	18.5	15.5
Treat. A	50	10	0.9	40.0
Treat. B	10	50	7.6	4.0
Treat. C	10	10	13.5	14.6
Treat. D	50	50	1.0	6.3

Table 5. Effects of varietal resistance on the equilibrium densities (per m^2) of GRL ($\overset{*}{X}$) and Lycosa ($\overset{*}{Y}$).

Resistance of rice plant[1)]	GRL ($\overset{*}{X}$)	Lycosa ($\overset{*}{Y}$)
0	15.5	18.5
10	11.3	15.9
20	7.3	12.8
30	4.7	10.3
35	3.8	9.2
38	extinct	extinct

[1)] Degree of resistance is expressed in terms of the percentage to which population growth (X_3/Y_1) is suppressed

Under natural conditions, it should be noted that neither the prey nor its predators reproduce during the winter, and that both of them are exposed to winter mortality before they resume their reproduction in the following spring. This means that the equilibrium point exists only in theory, the system starting each year from a combination of relatively low densities of both species. Another point which deserves to be mentioned is that the present system is constructed on the basis of empirical data. Consequently, the conclusions drawn only hold within the range of observed values and any extrapolation should be made with caution.

CONCLUDING REMARKS

Our systems models have been constructed almost independently, on their own empirical data. Hence, their synthesis in one, unified model remains to be achieved in the future. It is felt that the adoption of such concepts as incidence-days and physiological time, which can be measured by the law of thermal constant, is necessary to do so.

However, our systems study does suggest several points which should be taken into account in planning GRL-RDV management.

(1) Natural enemies are more vulnerable to insecticides than prey species because the insecticide affects them in two ways: by killing natural enemies directly and through inducing food shortage by killing their hosts (prey). On the other hand, an abundance of predators (*Lycosa*) improves the resilience (Holling, 1973) of the GRL population system, reducing the chances of GRL escaping from predation (Kiritani, 1976).

(2) Spiders alone, although they contribute to the suppression of the GRL population, are not effective in controlling RDV. A feasible means of controlling RDV, however, is suggested by the integrated use of *Lycosa* with a moderately resistant variety of rice. Accordingly, more effort should be made to breed resistant *japonica* varieties to RDV or GRL.

(3) It is suggested that cultural practices, especially winter ploughing of fallow paddy fields, is effective in suppressing RDV prevalence. This has already been demonstrated on a practical scale (Nakasuji and Kiritani, 1976).

REFERENCES

Holling, C.S. (1973), Resilience and stability of ecological systems, *Ann. Rev. Ecology and Systematics*, 4, 1-23.

Ishii, M., S. Yasuo, and T. Yamaguchi (1970), Epidemiological studies on rice dwarf disease in Kanto-Tosan district, Japan, *J. Central Agric. Expt. Sta.*, 14, 1-115.

Kassai, T. and K. Ozaki (1972), Injury of the rice plant at the full heading stage caused by sucking of the green rice leafhopper adults, *Proc. Assoc. Plant Prot. Shikoku*, 7, 1-4.

Kawahara, S. (1975), Population dynamics of micryphantid spiders in the paddy field, *Bull. Kochi Inst. Agr. & Forest Sci.*, 7, 53-64.

Kawahara, S., K. Kiritani, and T. Sasaba (1971), The selective activity of rice-pest insecticides against the green rice leafhopper and spiders, *Botyu-Kagaku*, 36, 121-128.

Kawahara, S., K. Kiritani, and N. Kakiya (1974), Population biology of Lycosa pseudoannulata, *Bull. Kochi Inst. Agr. & Forest Sci.*, 6, 7-22.

Kiritani, K. (1977), Systems approach for management of rice pests, *Proc. 14th Int. Congr. Entomol.*, Washington, D.C., in press.

Kiritani, K. and J.P. Dempster (1973), Quantitative evaluation of natural enemy effectiveness, *J. Appl. Ecol.*, 10, 323-330.

Kiritani, K. and S. Kawahara (1973), Food-chain toxicity of granular formulations of insecticides to a predator, Lycosa pseudoannulata of Nephotettix cincticeps, *Botyu-Kagaku*, 38, 69-75.

Kiritani, K. and N. Kakiya (1975), An analysis of the predator-prey system in the paddy field, *Res. Popul. Ecol.*, 17, 29-38.

Kiritani, K., et al. (1970), Studies on population dynamics of the green rice leafhopper, Nephotettix cincticeps: Regulatory mechanism of the population density, *Res. Popul. Ecol.*, 12, 137-153.

Kiritani, K., et al. (1972), Quantitative evaluation of predation by spiders on the green rice leafhopper, Nephotettix cincticeps, by a sight-count method, *Res. Popul. Ecol.*, 13, 187-200.

Lotka, A.J. (1925), *Elements of Physical Biology*, Williams & Wilkins, Baltimore, 465.

Miyai, S., K. Kiritani, and T. Sasaba, An empirical model of the predator-prey system in the paddy field, in preparation.

Nakasuji, F. (1974), Epidemiological study on rice dwarf virus transmitted by the green rice leafhopper, Nephotettix cincticeps., *JARQ*, 8, 84-91.

Nakasuji, F. and S. Nomura (1968), A study of the injury by the green rice leafhopper, *Proc. Assoc. Plant Prot. Shikoku*, 3, 21-26.

Nakasuji, F. and K. Kiritani (1971), Inter-generational changes in relative abundance of insects infected with rice dwarf virus in populations of Nephotettix cincticeps UHLER, *Appl. Ent. Zool.*, 6, 75-83.

Nakasuji, F. and K. Kiritani (1972), Descriptive models for the system of the natural spread of infection of rice dwarf virus by the green rice leafhopper, *Res. Popul. Ecol.*, 14, 18-35.

Nakasuji, F. and K. Kiritani (1976), Epidemiological analysis of the post-war RDV outbreaks in Japan, *Shokubutu Boeki*, 30, 48-52.

Nakasuji, F., K. Kiritani, and E. Tomida (1975), A computer simulation of the epidemiology of the rice dwarf virus, *Res. Popul. Ecol.*, 16, 245-251.

Rosenzweig, M.L. (1971), Paradox of enrichment: destabilization of exploitation ecosystems in ecological time, *Science*, 171, 385-387.

Sasaba, T. (1974), Computer simulation studies on the life system of the green rice leafhopper, Nephotettix cincticeps UHLER, *Rev. Plant Prot. Res.*, 7, 81-98.

Sasaba, T. and K. Kiritani (1975), A system model and computer simulation of the green rice leafhopper populations in control programmes, *Res. Popul. Ecol.*, 16, 231-244.

Sasaba, T., K. Kiritani and T. Urabe (1973), A preliminary model to simulate the effect of insecticides on a spider-leafhopper system in the paddy field, *Res. Popul. Ecol.*, 15, 9-22.

Sugino, T. (1975), Economic injury level of rice insect pests, *Shokubutu Boeki*, 29, 263-267.

Takahashi, Y. and K. Kiritani (1973), The selective toxicity of insecticides against insect pests of rice and their natural enemies, *Appl. Ent. Zool.*, 8, 220-226.

Volterra, V. (1926), Variazioni e fluttuazioni del numero d'individui in specie animali conviventi, *Mem. R. Accad. Naz. dei Lincei, Ser. 6*, 2, 31-113.

Table 4. Effect of weekly insecticidal treatments on the *Lycosa* (spider)-*Nephotettix* (leafhopper) system (computer simulation).

	Effectiveness (% kill)		Equilibrium density (m^2)	
Treatment	Lycosa	Nephotettix	Lycosa	Nephotettix
Control	0	0	18.5	15.5
Treat. A	50	10	0.9	40.0
Treat. B	10	50	7.6	4.0
Treat. C	10	10	13.5	14.6
Treat. D	50	50	1.0	6.3

Table 5. Effects of varietal resistance on the equilibrium densities (per m^2) of GRL ($\overset{*}{X}$) and Lycosa ($\overset{*}{Y}$).

Resistance of rice plant[1)]	GRL ($\overset{*}{X}$)	Lycosa ($\overset{*}{Y}$)
0	15.5	18.5
10	11.3	15.9
20	7.3	12.8
30	4.7	10.3
35	3.8	9.2
38	extinct	extinct

[1)] Degree of resistance is expressed in terms of the percentage to which population growth (X_3/Y_1) is suppressed

b) To develop alternative, ecologically desirable tactics for use in suppressing major pests (i.e., to reduce the use of broad spectrum biocides and lessen the adverse effects of their use).

c) To develop better methods of collecting, handling and interpreting relevant biological, meteorological and crop production data.

d) To develop the use of systems analysis and modelling as a central unifying research-guiding tool in the development of crop production and protection systems.

e) To develop models of the crop production and pest systems, integrate these with economic analysis (impact, etc.), and conduct pilot tests on the use of the combined management models for each crop system.

Below is a chart of the administrative structure which oversees the research in the five crops by the cooperating universities.

CHART OF ADMINISTRATIVE STRUCTURE

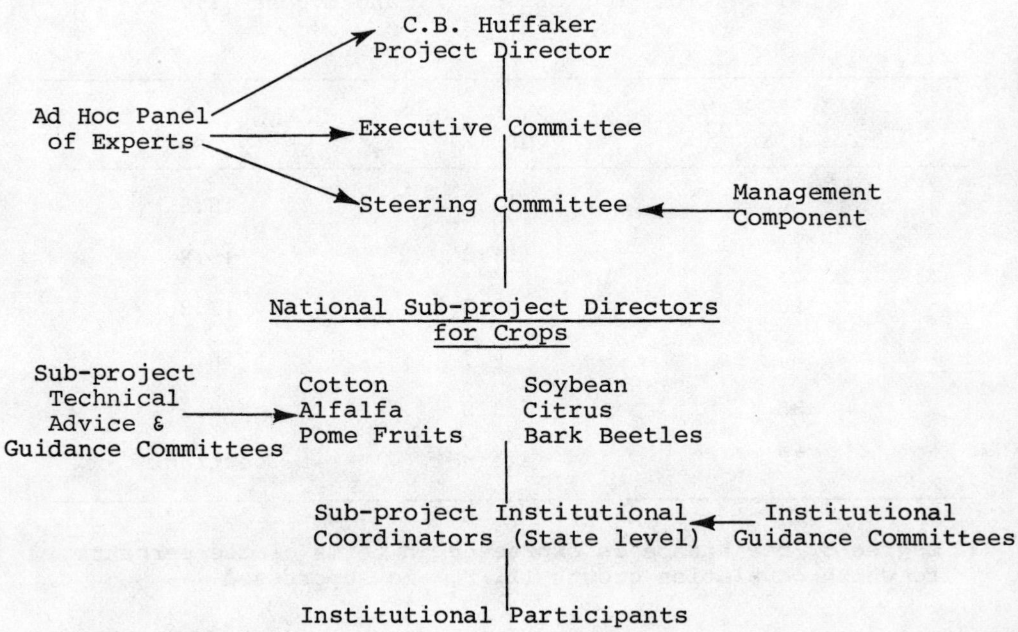

APPLIED POPULATION ECOLOGY: MODELS FOR CROP PRODUCTION AND PEST MANAGEMENT

A. P. Gutierrez[1] and Y. Wang[2]

INTRODUCTION

Pest management is a catchy term used by entomologists who naturally focus on the invertebrate pest, but it may be a misleading term. The experience of the Integrated Pest Management Projects (NSF/IPM) in the United States has shown that what entomologists really mean when they discuss pest management is crop production and protection, and that pest management is but one aspect; like plant nutrition or some other discipline.

"Systems analysis" has a considerable contribution to make to crop production, provided the essence of the crop production problems are distilled out in a mathematically tractable form. The methods of systems analysis are quite diverse (Watt 1966), but whichever method is chosen, it must be adequate for dealing with problems in population ecology: populations of plants, pests and natural enemies, as modified by weather and by man's various agronomic inputs. Figure 1 depicts a very trivial representation of this complex problem from the farmer's point of view. Because crops are grown for profit, the output is necessarily a monetary one.

If the crop production system is to be analyzed, it must first be put into some conceptual framework. Given that this framework exists, the results of the analysis (i.e., understanding) can then be used to develop optimization strategies for pest control, plant breeding, or some other aspect of crop production. This chapter shows how the concepts of population ecology can be applied to the study of crop production and pest management problems. While the emphasis here is at the population level, the models presented below may be complicated with relative ease with the appropriate and necessary biological variables to make them realistic: the models are, after all, designed to examine field problems, not esoteric ones. The mathematics, while difficult for biologists, are trivial by mathematical standards.

---

[1]Associate Entomologist and [2]Specialist in Mathematics, Division of Biological Control, University of California, Berkeley.

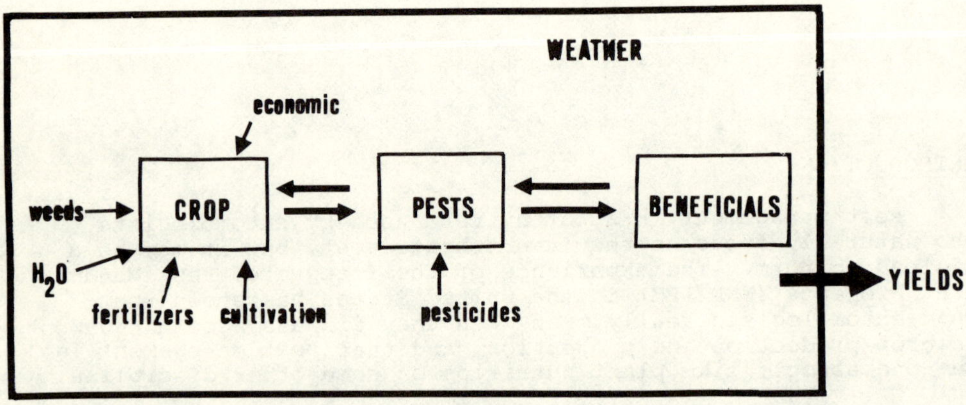

Figure 1. A simplified representation of an agroecosystem.

THE CONCEPTUAL FRAMEWORK

The plant ecologists, Harper and White (1974), have stated that populations of plants are composed of individuals of different ages, and that each plant is further composed of populations of plant parts (e.g., root, stem, leaf and fruit tissues) also of different ages. Our terminology is different from theirs, but the concept remains valid. In a crop, the ages of individual plants is more uniform, but the second supposition holds.

The population processes in plants appear to be analogous to those found in animal populations. The production of new photosynthetic material by the plant and its allocation to leaves, stems and roots (growth) or fruit parts (births) are, in all respects, analogous to the growth and birth processes in animal populations. The plant growths by trapping solar energy via photosynthesis, while all higher trophic levels use this trapped energy in lower ones until all of the energy is dissipated (i.e., via entropy). The death processes are similarly analogous. Whole plants or plant parts may die. In the latter case, the parts die when leaves fall off or stem and root tissues become woody, or they may be killed by predators, which at this level of abstraction we call herbivores for nomenclatural convenience. Plants also exhibit density-dependent responses. For example, crowding produces smaller, less hardy and less fecund individuals just like animals do.

Annual cotton (Gossypium hirsutim L.) turns out to be a very good experimental plant species. Gutierrez et al. (1975)

describe a population model for cotton growth and development, which conceptually was derived from earlier work on insect populations (Gutierrez et al., 1974). The notions concerning the birth and death processes in Acala SJ-11 cotton which determine the dynamics of the crop were presented there in a less vigorous form. Wang et al. (1977) put this model into an analytic form using demographic techniques common to animal ecology (von Foerster, 1959; Sinko and Streifer, 1967). This work is summarized here. Conceptually, the same processes apply equally well to perennials like alfalfa (Gutierrez, et al., 1976).

THE MATHEMATICAL FRAMEWORK

The classic population models (e.g., Thompson, 1924; Nicholson-Bailey, 1935; Lotka, 1925; Volterra, 1931; etc.), ignored age structure, and considered all individuals equal. These models are not suitable for most field problems. Slobodkin (1954), Frank (1960) and Auslander et al. (1974) in laboratory studies, and in a field study, Hughes and Gilbert (1968) added other population variables, such as age structure, to their models. The Leslie Matrix (Leslie 1945) is a convenient mathematical form for examining population models which include age structure (i.e., time varying life-tables, Hughes 1963). The major conceptual weakness of the Leslie Matrix model is that it considers that all individuals in a cohort (i.e., those born at the same time) develop at the same rate.

The Leslie model can be represented as follows:

$$\begin{bmatrix} N_{o,t+\Delta t} \\ N_{1,t+\Delta t} \\ N_{2,t+\Delta t} \\ \vdots \\ N_{n,t+\Delta t} \end{bmatrix} = \begin{bmatrix} mx_o(\cdot) & mx_1(\cdot) & mx_2(\cdot) & \cdots & mx_n(\cdot) \\ lx_o(\cdot) & & & & \vdots \\ 0 & lx_1(\cdot) & & & \vdots \\ & & lx_2(\cdot) & \ddots & \vdots \\ 0 & & & lx_{n-1}(\cdot) & 0 \end{bmatrix} \begin{bmatrix} N_{o,t} \\ N_{1,t} \\ N_{2,t} \\ \vdots \\ N_{n,t} \end{bmatrix}$$

where the age dependent birth $(mx_i(\cdot))$ and survivorship $(\ell x_i(\cdot))$ rates are complex functions incorporating much of the complexity described above. For example: $\ell x_{i,t}(\cdot) = f(\text{age, nutrition, crowding, predation, } \ldots .)$, and is the product of the survivorship probabilities of the k factors (i.e., $\prod_{j=1}^{k} \ell x_{i,j} =$

$\ell x_{1,j} \cdot \ell x_{2,j} \cdot \ell x_{3,j} \cdot \ . \ . \ \ell x_{n,j})$. $mx_{i,t}(\cdot)$ = f (age, nutrition, crowding, predation, . . .), and in this case the factors (e.g., age) scale the potential birth rate from some maximum.

A continuous form of this discrete model is the von Foerster model (von Foerster, 1959), which was originally developed to describe the growth of cell populations.

$$\frac{\partial N}{\partial t} + \frac{\partial N}{\partial a} = - \mu(\cdot) \cdot N(t,a) \quad .$$

In this model, N(t,a) = the number density function; it could be mass or energy units, t = time, a = age and $\mu(\cdot)$ = a multi-factor death function incorporating ℓx and mx factors. The model requires two conditions, N(O,a) is the initial age distribution and N(t,0) is the density of newborn. This model can either be solved numerically or it can be solved analytically (see Wang et al., 1977).

This model can be generalized to several variables (see Streifer, 1975, work by Frank, 1960, on *Daphnia pulex*, and Slobodkin, 1954, on *Daphnia obtusa*). Slobodkin's model is of particular interest because, in addition to age and time, he considered that size should be taken into account in defining classes of physiologically identical organisms. The specific merits or problems of his discrete model will not be discussed here.

Sinko and Streifer (1967), and later, Streifer (1975), proposed a continuous form of a general model, several variations of which we have adopted. Let N(t,a,m) be the population density function depending on t,a, and m = mass or size. Many biological attributes of a population at any point in time are determined by N(t,a,m). For example, the total number of organisms between ages (a_o,a_1) and mass (m_o,m_1).

$$\int_{a_o}^{a_1} \int_{m_o}^{m_1} N(t,a,m)\,dm\,da \quad ,$$

while the total biomass of all organisms at time t is

$$\int_{o}^{\infty} \int_{o}^{\infty} mN(t,a,m)\,dm\,da \quad .$$

If we assume that dt = da, then

$$\frac{\partial N}{\partial t} + \frac{\partial N}{\partial a} + \frac{\partial}{m} g(\cdot)N(t,a,m) = -\mu(\cdot)N(t,a,m)$$

where g $(\cdot)$ and $\mu(\cdot)$ are the mass growth rate and death rates respectively, which are specific to the problem in hand. The equation is not limited to these variables. In general, suppose that there are k physical characteristics, $m_1, m_2, \ldots, m_k$, then the balance equation is

$$\frac{\partial N}{\partial t} + \frac{\partial N}{\partial a} + \sum_{j=1}^{k} \frac{\partial}{\partial m_j} (g_j(\cdot)N) = -\mu(\cdot)N$$

where $g_j = \frac{dm_j}{dt}$ = the growth rate of characteristic m_j.

We are now ready to discuss the applicability of this simulation model to modelling trophic levels. Modelling of all trophic levels has three distinct but essential parts: (a) developing the analytical framework, (b) collection of field and laboratory data and (c) model testing against data sets not used in the construction of the model. The aim of simulation model building should first be understanding and not merely prediction. Figures 2 and 3 depict some of the components of the alfalfa and cotton agroecosystem.

THE COTTON ECOSYSTEM MODEL

Plant Population Model

A recent book by Evans (1975) discusses the general state of development of the older plant modelling and crop physiology literature. In recent years, several workers have developed models describing the growth and development of single plants (de Wit et al., 1970; Hesketh et al., 1971; Duncan et al., 1971; Hesketh et al., 1972; McKinion et al., 1974; Stapleton et al., 1973; and Fick and Loomis, 1974). For the most part, these models are complex algorithms. Some of them have been good simulators, but have been less than satisfactory for use in agroecosystem studies. The emphasis has been placed more on physiological aspects than on functional population dynamics. A more critical shortcoming is that in most cases these models lack a definable mathematical form which greatly hinders their usefulness. Other models of a more analytical nature (McKinion et al., 1975) distill much of the essence of the problem, but are unrealistically structured (e.g., they ignore age structure).

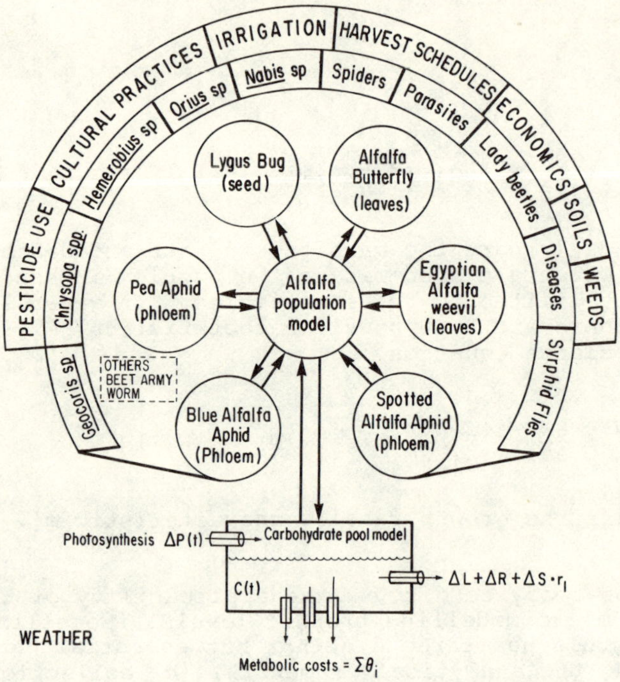

Figure 2. Some of the potential interactions in an alfalfa ecosystem. Notice that all interactions ultimately impinge on the carbohydrate pool model.

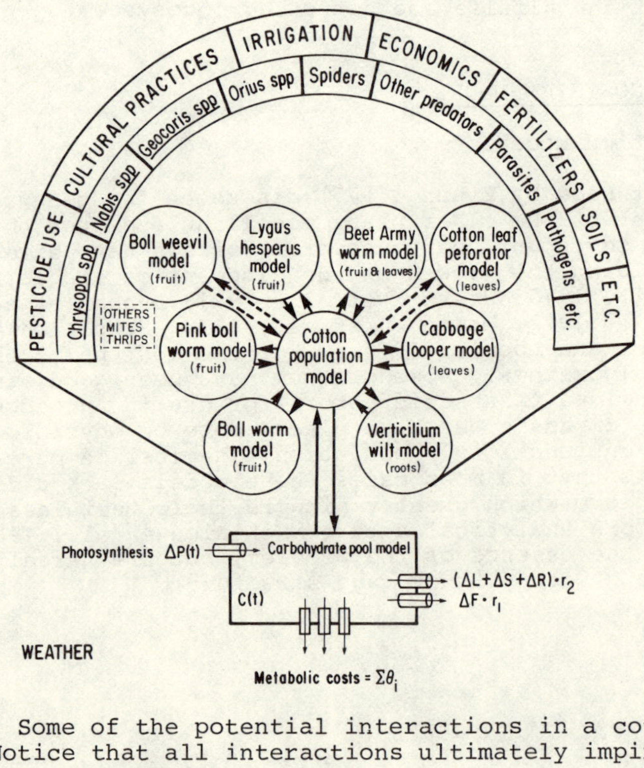

Figure 3. Some of the potential interactions in a cotton ecosystem. Notice that all interactions ultimately impinge on the carbohydrate pool model.

The model of McKinion et al. is basically of the following form:

$$\frac{dW}{dt} = dP/dt - G_r \cdot dQ/dt - R_o W - R_s \cdot dC/dt \ .$$

W = total plant weight (plant weight (Q) + carbohydrate reserves (C))

P = photosynthate

G_r = growth respiration

R_o = maintenance respiration

R_s = respiration associated with stored starch

t = time in day degrees

Miles et al. (1973) developed a similar model for alfalfa growth and development, but it was less analytical in nature.

Briefly, the models (Gutierrez et al., 1975 [discrete form] Wang et al., 1977 [continuous form]) are composed of balance equations for plant, leaf, stem, root, fruit populations and the nutrient pool (= photosynthate). The equations and initial conditions for these models can be written as

$$\frac{\partial \rho}{\partial t} + \frac{\partial \rho}{\partial a} = - \mu_\rho(\cdot) \cdot \rho(t,a) + I_\rho \qquad \begin{aligned} \rho(t,o) &= \beta_\rho(t) \\ \rho(0,a) &= \gamma_\rho(a) \end{aligned} \tag{1}$$

$$\frac{\partial L}{\partial t} + \frac{\partial L}{\partial a} = - \mu_L(\cdot) \cdot L(t,a) \qquad \begin{aligned} L(t,o) &= g_L(t) \\ L(o,a) &= \alpha_L(a) \end{aligned} \tag{2}$$

$$\frac{\partial S}{\partial t} + \frac{\partial S}{\partial a} = - \mu_S(\cdot) \cdot S(t,a) \qquad \begin{aligned} S(t,o) &= g_S(t) \\ S(o,a) &= \alpha_S(a) \end{aligned} \tag{3}$$

$$\frac{\partial R}{\partial t} + \frac{\partial R}{\partial a} = - \mu_R(\cdot) \cdot R(t,a) \qquad \begin{aligned} R(t,o) &= g_R(t) \\ R(o,a) &= \alpha_R(a) \end{aligned} \tag{4}$$

$$\frac{\partial F}{\partial t} + \frac{\partial F}{\partial a} = - \mu_F(\cdot) \cdot F(t,a) \qquad \begin{aligned} F(t,o) &= \beta_F(t) \\ F(o,a) &= \gamma_F(a) \end{aligned} \tag{5}$$

$$\frac{\partial M}{\partial t} + \frac{\partial M}{\partial a} = - \mu_M(\cdot) \cdot M(t,a) \qquad \begin{aligned} M(t,o) &= g_M(t) \\ M(o,a) &= \alpha_M(a) \end{aligned} \tag{6}$$

where the independent variable t and a are time and age respectively measured in day degrees (D^o). Note that in its simplest deterministic form dt = da. In fact, dt ≠ da in nature because individuals in the population tend to age at different rates. L(t,a), S(t,a), R(t,a) and M(t,a) are mass density functions for leaf, stem, root and fruit tissues respectively, while $\rho(t,a)$ and F(t,a) are number density functions for whole plants and fruit. The term I (Equation 1) is a net immigration rate (in a general sense, it could be new seeds or other plant propagation methods). This term is not applicable to domestic cotton crops, but is important for wild cotton where new seed may be blown into the area encompassing the study population. It is included in Equation 1 to make the model general for any population of plants. The various $\mu(\cdot)$ are complex death rate functions. The model is basically a canopy model encompassing a population of plants, but if $\rho=1$ in Equation 1, the model collapses to a single plant model. All of these models are coupled via the carbonhydrate pool model.

The carbohydrate pool sub-model differs radically in form (Figure 4).

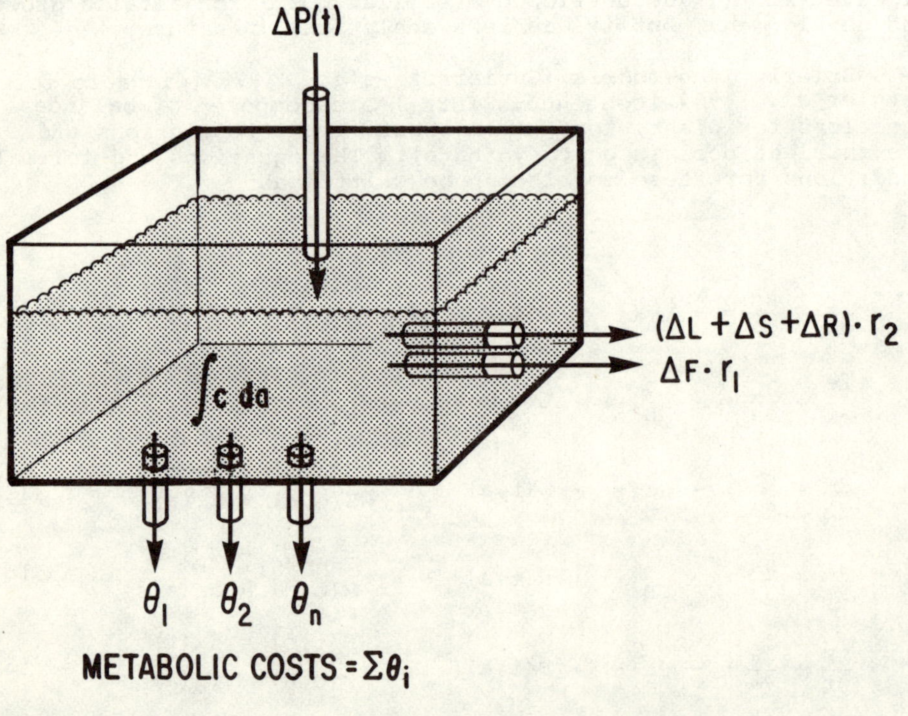

Figure 4. The metabolic pool model for allocating photosynthate (P), carbohydrate reserves (C) to meet metabolic needs (θ_i) of the plant or growth of plant parts (e.g., leaves (L), stem (S), root (R), or fruit (F). Note that the levels of the outflows indicate a priority scheme, and r, and r_2 indicates the fraction of the maximum growth achieved during some $\Delta t(0 \leq r_i \leq 1)$.

The equations for estimating photosynthesis used in the model are derived from McKinion et al. (1974). Let $\emptyset(t)$ be the carbohydrate present in the pool at time t. Then $\emptyset$ satisfies the following equation.

$$\frac{d\emptyset}{dt} = \frac{dP}{dt} - \left(\sum_{i=1}^{3} \theta_i(t) + \dot{R} + \dot{S} + \dot{M} + \dot{L}\right) .$$

If $d\emptyset/dt$ is approximately by $(-\emptyset(t)+\emptyset(t+\Delta t))/\Delta t$, the difference representation of Equation (3) is then

$$\emptyset(t+\Delta t) = \emptyset(t) + \frac{dP}{dt} \cdot \Delta t - \left(\sum_{i=1}^{3} \theta_i(t) + \dot{R} + \dot{S} + \dot{M} + \dot{L}\right) \cdot \Delta t .$$

dP/dt is the rate the photosynthate is produced, $\emptyset(t)$ is the reserve carbohydrate available, the θ_i's are the metabolic costs associated with plant growth and development (C.F. McKinion et al. 1974), and $\dot{R}$, $\dot{S}$, $\dot{M}$ and $\dot{L}$ are the realized plant growth rates. The cotton growth is determined principally by photosynthate availability (= supply (Q)) to meet daily age dependent demands (= D) by the various plant parts (L, S, R, M) and the θ_i's.

The computation of Q, D, as well as the priority scheme for allocating dry matter can be described algebraically as follows, while the process can be visualized as shown in Figure 4. The levels of the taps for dry matter allocation can be viewed as a priority scheme.

The total demand (D) at time t is given by:

$D = \sum^{3} \Delta\theta_i = \Delta L + \Delta R + \Delta S + \Delta M$, where the $\Delta\theta_i$ are respiratory losses and R, S, L and M are the maximum growth rates by the plant parts (roots, stems, leaves and fruit respectively). If $D \leq Q_t$ (i.e., $r = Q/D > 1$), then

$\Delta C_t = Q_t - D_t$, otherwise, the following priority scheme occurs.

1. $Q_1 = Q_t - \sum_{i=1}^{2} \Delta\theta_1$, where $\theta_3 \cdot \Delta t = (\theta_M + \theta_L + \theta_S + \theta_R)\ \Delta t$

2. $r_1 = Q_1/(M + \theta_M) \cdot \Delta t$, where $0 \leq r_1 \leq 1$, $Q_1 > 0$ and $\overset{*}{\theta}_3 = \theta_M$

3. $Q_2 = Q_1 - (\theta_M + M) \cdot \Delta t \cdot r_1$ (i.e., for fruit = M)

4. If $Q_2 > 0$, $r_2 = Q_2/(S + R + L + \overset{*}{\theta}_3)\ \Delta t$, for $Q_2 > 0$ and $0 \leq r_2 \leq 1$

5. $\Delta C_t = Q_2 - [\overset{*}{\theta}_3 + S + R + L] \cdot \Delta t \cdot r_2$ (i.e., for all other plant parts).

The second $0 \leq Q/D = r_2 < 1$ ratio is used to scale the rates of mainstem node and fruit point production, as well as to estimate the rate and ages of fruit (squares and small bolls) to be shed.

The time when $r < 1$ occurs may vary depending upon several factors (e.g., plant density, weather, etc.), but its impact can be readily observed in the field as slower dry matter accumulation in leaves, stems and roots, reduced mainstem node and fruit point production, the occurrence of peak squaring and rapid growth of large bolls (see Figure 5). The model is a good simulator of cotton growth and development in the San Joaquin Valley. The parameter modifications required to convert the Acala model to simulate the Delta pine variety of cotton grown in Arizona are given by Gutierrez, Butler et al. (1977).

Collection of Plant Data

The parameter estimates required to construct models for plant growth and development are of several kinds:

1. developmental phenology.
2. estimates of photosynthesis under various conditions of leaf age and light intensity.
3. estimates of dry matter partitioning amongst plant parts.
4. carbohydrate sink-source relationships.
5. density-dependant growth relationships.
6. mortality of plants and plant parts due to various causes.
7. the effects of weather on plant growth and development; especially temperature, light and moisture.
8. the effects of agronomic practices on plant growth and development.

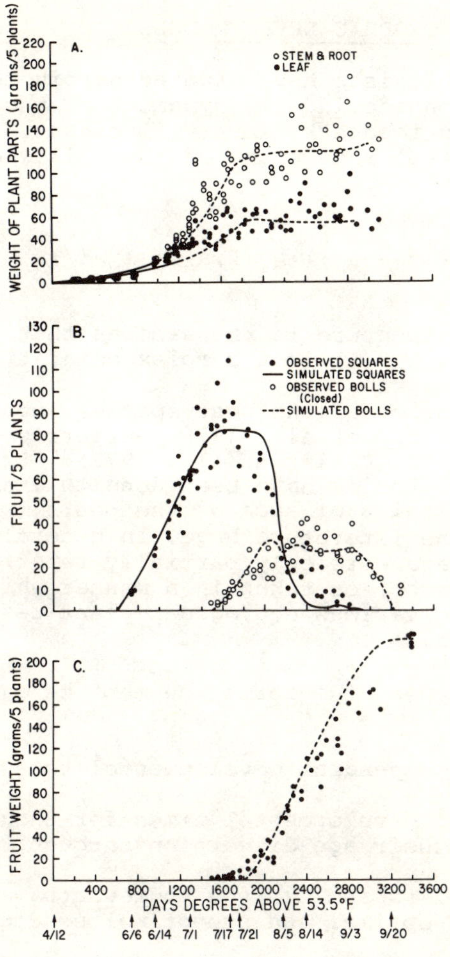

Figure 5. The phenology and development of a cotton crop in Corcoran, California during 1975: (A) dry matter accumulation in stems + roots and leaves, (B) the observed fruit patterns and (C) the accumulation of dry matter in fruit.

It is certain that sub-models for more accurately estimating photosynthetic efficiency (e.g., light penetration) and water use by plants will be forthcoming soon, and this model is formulated with these developments in mind. These recent advances are discussed in a later section.

HERBIVORE AND HIGHER TROPHIC LEVEL MODELS

In the past, zoologists have studied herbivore populations without adequately considering the dynamics of their food. As we shall see, this feedback is very important. The herbivore and higher trophic level models are of the following form.

$$\frac{\partial N_i}{\partial t} + \frac{\partial N_i}{\partial a} = - \mu(\cdot) \cdot N_i(t,a) + I_{N_i}(\cdot) \ .$$

Note that the model structure is the same as that for modelling the plant except that $I_{N_i}(\cdot)$ is a complex emigration or immigration rate function of the pest species into or out of the field (C.F. Gutierrez et al., 1976; Gutierrez, Leigh et al., 1977; Gutierrez, Butler et al., 1977). Extensive field and laboratory studies have been done to examine the species specific biologies of each of the pests (Figure 2). Relevant data from the literature have, in general, been sparse because the data were only partially reported, it was inconsistent or in error, or taken in a manner which was not useful for developing various ecological plant-pest-predator-weather relationships.

The following types of information must be gathered for insect species:

1. temperature dependent developmental rates.
2. estimates of developmental times for each life stage, as well as their age dispersion through time.
3. estimates of ℓx and mx (see Andrewartha and Birch, 1954) under optimal and suboptimal conditions.
4. data on their seasonal phenology, population dynamics and age structure.
5. overwintering characteristics.
6. site and age preference for plant parts.
7. feeding rates.
8. sex ratios.
9. detailed observation on behavior, etc.
10. detailed field life-table data for use in developing various mortality sub-models (see next section).

Population models for several of the important pests of cotton have been developed, and the computer algorithms are, by and large, discrete approximations of the models proposed here. A partial list of these models is:

1. Beet armyworm - cabbage lopper/cotton - Gutierrez et al. (1975), Hogg and Gutierrez (in progress).

2. Bollworm/cotton - *Stinner et al. (1974), Wilson and Gutierrez (in progress), L. Brown et al. (in progress).

3. Boll weevil/cotton - Wang et al. (1977), *Jones et al. (1974).

4. *Lygus hesperus*/cotton - Gutierrez, Leigh et al. (1977).

5. Pink bollworm/cotton - Gutierrez, Butler et al. (1977).

A model for mite dynamics on cotton is currently in progress (J. Carey, in progress).

Coupling Trophic Level Models

Coupling of plant herbivore activities is not conceptually too difficult, given that the appropriate data are well in hand (see above). First, we must recognize in coupling the two models that the time step for the plant may not equal the time step for the herbivore(s). The developmental rate of the herbivore life stages may also vary or they may be influenced by the age or quality of the plant part attacked (e.g., pink bollworm - Gutierrez, Butler et al., 1977). Next, we must integrate age preferences for plant part, searching success, hunger, attack rates, etc., in a predation model for each pest.

A fundamental understanding of the attack processes of a pest on a crop is very germane to the development of sound economic thresholds or to formulate economic models to assess the damage (Na = numbers or mass attacked) of the pest on a wide scale. For plants, Na may be fruit numbers, leaf, stem, fruit or root mass, photosynthate and/or wound healing losses. Figure 4 depicts the nature of the problem, as the plant's ability to compensate for damage is to a large degree dependent

*Those preceded by an asterisk have not been successfully coupled with a suitable cotton plant model.

on available reserves. If $\int cda$ is the total reserves, $D_{max} = \Delta L + \Delta S + \Delta R + \Delta M$ = the maximum demand for plant growth, ΔP is the new photosynthate and $\sum_i \theta$ are the various metabolic costs, then reserves are utilized when $D_{max} + \sum_i \theta > \Delta P$. If maximum growth is to occur, the demand $\leq$ supply of photosynthate

$$D_{max} + \sum \theta_i \leq P + \lambda \int cda \quad ,$$

where λ is some rate of reserve availability. It is possible that under stress conditions (e.g., insect damage = Na) the ability of the plant to compensate is severely impaired. Reduced yields can occur because of herbivore injury (Na), reduced demand (D^*_{max}) or reduced photosynthesis ($\overset{*}{P}$).

$$Na + D^*_{max} + \sum \theta_i > P^* + \lambda \int cda \quad .$$

A similar model exists for herbivores and also predators - Figure 6, but the important component is hunger as it affects the predator's search behavior. The animal has some age dependent maximum gut capacity (Ω) which may be full or may be partially empty. If the gut is partially empty, the insect is hungry (i.e., it can eat), but the amount it eats (Na) is the mass of prey it captures. The model for hunger is:

$$h_t = h_{t-1} - Na_t + Mx_t + g_t + \Delta\Omega_t + \sum \psi_{i,t} \quad ,$$

where

h = the mass of prey the animal can eat at time t ,
Na = the mass of prey captured ,
Mx = the mass of progeny produced ,
g = the mass growth of the animal ,
$\sum \psi_i$ = metabolic costs .

The priority of nutrients for g and Mx change with age and h (e.g., adults stop growing but do produce progeny), while survivorship (ℓx) is also a function of h (e.g., Gutierrez et al., 1976).

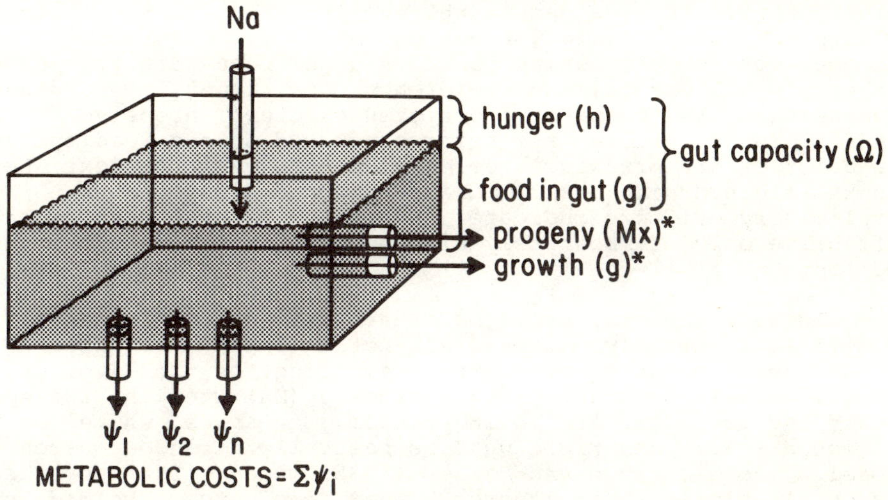

Figure 6. The metabolic pool model for herbivores and predators. Na is the mass prey captured. The levels of the outflows indicate priority levels which depend on age and hunger.

"PREDATION" MODELS OR ESTIMATING ATTACK SUCCESS

Herbivore/Predation

An herbivore is a plant predator, hence the relevant predation models apply equally well to them as to the higher trophic levels. If an herbivore attacks a population of leaves, it alters their age structure (by consuming mass), and causes wound healing losses. Because individual leaves may be only partially eaten, and still live, mass rather than numbers are the units of attack (Na). Many herbivores attack fruit and cause them to abscise or be unavailable for other herbivores. In such cases, Na will refer to numbers, though its relationship to mass may also be useful.

Cotton and alfalfa share many of the same predators of their pests as well as some pests. The principal general entomophagous predators are *Geocoris pallens* Stal, G. *punctipes* Say, *Chrysopa carnea* (Stephens), *Nabis americoferus* Carayon and *Orius tristicolor* (White). The migration of insect pests from alfalfa to other crops is well documented (e.g., lygus bug--Sevacharian and Stern, 1972), but the movement of the natural enemies from alfalfa is less well understood (Ehler et al., 1974 and Eveleens et al., 1974).

Predators of plants or animals may be polyphagous or oligophagus. In some cases, the predators may feed on both plants and animals (e.g., Lygus and Orius). The problem becomes more complex as we consider other behavioral components like prey preference, age preferences, switching, age dependent searching rates, predator interference, etc. Intensive field and laboratory studies are being conducted on the feeding behavior of all predator species listed above in cotton. Less is known concerning insect parasitoids and disease, but preliminary evidence indicates that these natural enemies are effective only at moderately high host densities; at least in cotton.

These behavioral problems create difficulties in developing models to accurately estimate Na. Lotka (1925) and Volterra (1931) describe an equation for estimating the number of prey (e.g., animals or plant parts) attacked (Na) when the two species (prey and predator) are moving randomly in space, while Thompson (1924) described what is basically a random search model. The Nicholson and Bailey (1935) model, like the others, is overly simple as it ignores hunger, age, etc. The models proposed by Holling (1959, 1966) were adequate to describe the activities of a single predator on a population of prey, but not for a population of predators on a population of prey, which is what we see in nature. Griffith and Holling (1969) attempted to expand this model to multiple predators, but it still proves inadequate.

Other models by Royama (1970), Rogers and Hassell (1974), and others, consider non-random distribution of prey and predator searching. While these models add complexity, they do not provide adequate estimates for Na.

Frazer and Gilbert (1976) and Gilbert et al. (1976) derive a more tractable model which overcomes most of the previous difficulties.

$$Na = N_o(1-e^{-(bTP/N_o)(1-e^{-aN_o/b})})/A$$

where

Na = the number of prey attacked at time t
N_o = the number prey present
b = the predator demand rate
a = the predator search rate
T = Δt
P = predator numbers
A = area to be searched.

These authors incorporated Holling's hunger component, but not on an individual predator level, rather, averaged over the predator population. This model was used with great success by Frazer and Gilbert to model coccinellid predation at low aphid densities, and by Wang et al. (1977) to model boll weevil (an herbivore) attack on cotton squares using Nicaraguan cotton data. This model provides more adequate estimates of Na at all host-predator densities, and estimates survivorship from predation (ℓx_s) in the large population models. The model needs to be expanded to incorporate relevant stochastic processes such as predator search rates, attack success, hunger, etc. This model has many interesting properties. For example,

as N_o becomes small, the term $1-e^{-aN_{o/b}} \rightarrow aN_{o/b}$ as $aN_{o/b} \rightarrow 0$,

or

$$1-e^{-aN_{o/b}} = 1 - 1 + aN_{o/b} - \frac{(aN_{o/b})^2}{2!} + \frac{(aN_{o/b})^3}{3!} - \cdot \cdot \cdot \cdot - \frac{(aN_{o/b})^{n-1}}{n-1!} + \frac{(aN_{o/b})^n}{n!}$$

which, when substituted into the model, yields

$$Na = N_o(1-e^{-aTP})$$

as $P \rightarrow 0$, the Nicholson-Bailey model becomes a Lotka-Volterra model because

$$1-e^{-aTP} = 1-1 + aTP - \frac{(aTP)^2}{2!} + \frac{(aTP)^3}{3!} - \cdot \cdot \cdot \cdot - \frac{(aTP)^{n-1}}{(n-1)!} + \frac{(aTP)^n}{n!} ,$$

and the model reduces to

$$Na = N_o \, aTP \quad .$$

The definitions of the parameters a and b are important to the model, and are described below.

a is derived from the Nicholson-Bailey model and equals

$$-\ln\left(\frac{N_o-Na}{N_o}\right) P/T/A \quad .$$

If P=1, T=Δt=1 and A=1 (i.e., standard conditions for estimating a empirically), then

$$a = -\ln\left(\frac{N_o-Na}{N_o}\right) = -\ln\left(1 - \frac{Na}{N_o}\right) = -\ln(1x) \quad .$$

The Nicholson-Bailey model is negative log survivorship, while the Lotka-Volterra model equals the mortality rate (i.e., effective area of search).

To estimate a empirically (for each species and life stage), a realistic field or laboratory experimental arena must be devised, and values of A and T assigned such that the behavior which underlies a is not greatly influenced (e.g., large changes in hunger ($\bar{h}$)) (see Frazer and Gilbert, 1976).

A general model following work of Frazer and Gilbert for multiple predator attack on several species (also several ages) of predators has been formulated by Jones, Wang and Gutierrez (unpublished).

Biophysical Models

Accuracy and further advancement of population modelling requires that relevant stochastic processes (e.g., Barr, 1973) and appropriate behavior be included, and that the underlying physiological processes affecting age-specific survivorship ($\ell x(\cdot)$) and natality ($Mx(\cdot)$) be described in rigorous mathematical terms. Considerable progress has been made in this area.

The basic premise in much of the work is that survival and developmental rates, and most other factors of plant and insect population dynamics are controlled by the thermal environment. For example, much of the earlier work on developmental rates of insects and plants used the notion of day degrees. This concept assumes that their developmental rates are exactly proportional to temperature throughout much of the range favorable to their environment (see Campbell et al., 1974, for a recent review). In fact, many studies have shown that this is not the case, and the extremes of the rate of development curve shown in Figure 7 depicts these departures from the linear model (Stinner et al., 1974). Sharpe and DeMichele (1977) developed a stochastic, thermodynamic model for the development of poikilotherm organisms. The model is derived from the Erying equation which assumes multiple activity states of the underlying developmental control enzymes. This model describes the entire relationship from a more theoretical and scientific base, and in addition, contributes to scientific understanding. As a result, the model allows for more accurate prediction of population phenology and ageing. An extension of this model by Sharpe et al. (unpublished) shows that the skew in the distribution in developmental times observed by Stinner et al. (1975) and others, results naturally from the transformation from developmental rates to emergence times. The underlying assumption is that the concentration of enzymes, which are rate controlling for development, are symmetrically distributed about some genetically determined mean concentration. These two simple, yet elegant, biophysical models are used to describe not only the rate of development of a population, but also the distribution of developmental times. Both of these models are easily incorporated in the population models developed above.

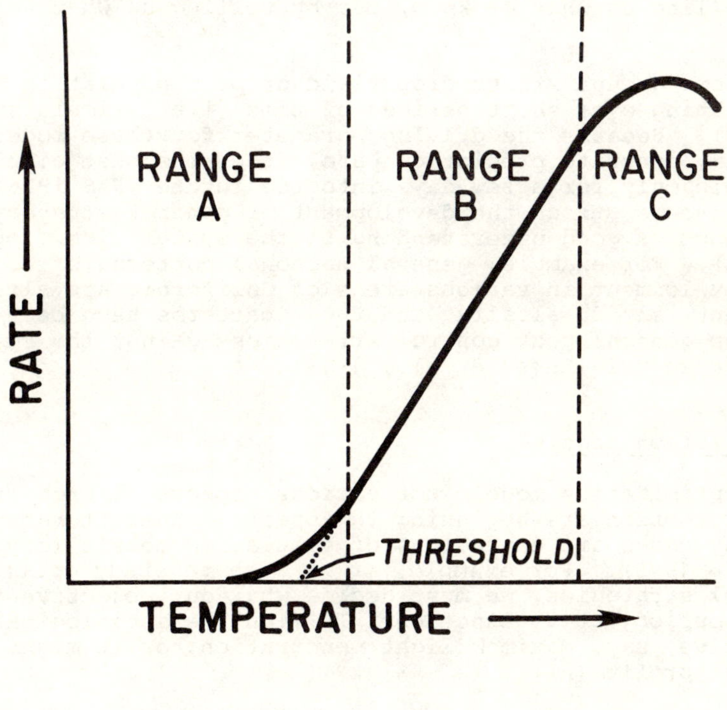

Figure 7. The relationship between the rate of development of poikilothermic organisms and temperature. In temperature ranges A are non-linear because of subleathal low and high temperature effects. The threshold (---) or T* is a reasonable approximation for the lowest temperature where the development of the organism can occur.

WHAT USE ARE MODELS IN COTTON PRODUCTION AND PLANT PROTECTION?

Probably the most important use of a model is that it forces the researcher to specify the components. Because of this, the process enables the worker to quickly define what is known about the problem and, as such, is useful in guiding the research.

Most of the above models are designed for use in developing crop production-protection strategies and hence must be examined for accuracy in the field under a variety of conditions. In a practical sense, the only check of a model's "validity" is its explanatory value for check data not used in the construction of the model (i.e., independent tests), but the test must include a wide range of variables. For example, the cotton models (Gutierrez et al., 1975) and Wang et al., 1977, see Figure 5) both require that the model mimic phenology of crop growth and development, dry matter accumulation in the various plant parts, and age structure of the fruit population. When discrepancies arise, the researcher must explain them, so the model may be immensely useful, not only for telling us what we know, but by telling us what we don't know.

Some groups stress crop yield or pest population trend forecasting over short periods of time (i.e., simulation models). Because the driving parameter for these models is weather, accurate predictions are, under the best circumstances, reliable only for a few days into the future (NAS 1976). Other groups stress the development of general strategy models based upon a good understanding of the system (i.e., optimization models). For example, general seasonal patterns of crop growth and development in various areas of California are already apparent, say in alfalfa, and these patterns have been used to develop general pest control strategies against the Egyptian alfalfa weevil (Regev et al., 1976).

Optimization Models

Optimization models for various aspects of crop production and protection are beginning to appear in the literature, and in many cases are extensions of simulation models (e.g., Regev et al., 1976). For example, if we wish to study optimal pest control strategies, we must decide what our objective is -- i.e., our objective function. It might be a biological objective, say, maximum light penetration, or it might be maximum profit (R).

Farmers usually want to maximize profit (R), hence their objective function is

$$\max R = B(x) - C(x)$$

where B(x) are all the benefits of using some decision variable (e.g., x_i = pesticides, biological control, etc.), while C(x) are all of the costs of using the procedures. The optimal level of pesticide application occurs when

$$\frac{dB(x)}{dx} = \frac{dC(x)}{dx} \quad .$$

The validity of this optimization result depends on the validity of the model used, as well as the validity of the questions asked of the model. Optimization models to establish sound economic thresholds and control strategies are currently in progress for pink bollworm (*Pectinophora gossypiella* (Saunders)) and *Lygus hesperus* Knight. Both of these species are key pests in various areas of the cotton belt.

The optimization problem may be more biological in nature (e.g., plant breeding). Crop genotypes are selected by man to meet specific needs. The standard plant breeding experiments usually involve planting many cultivars and assessing various attributes such as yield and quality at the end of the season. This is an arduous, time-consuming procedure. It is possible that a good weather-driven, plant model could help predetermine the characteristics crop plants should have for any particular area. The plant model tells us that the dry matter partitioning coefficient, the timing of fruit development, etc., determines to a large extent the growth and development of the crop (i.e., determine yields). The weather inputs (e.g., for Fresno, California) are part of the model, and characterize the region. The optimization model seeks to maximize yields (Y), hence the objective function is

$$\max Y = \max_{x_i,\ i=1,2,..,n} \int_0^{\infty}\int_0^{\infty} mF(t^*,a,m,x_1 x_2, \ldots, x_n, \cdot)\ da\ dm$$

where $F(t^*,a,m,x_1,x_2, \ldots, x_n, \cdot)$ satisfied the system of equations which describe the cotton model. Note that the x_i are decision variables (see below), m is the mass of fruit and F the number, t* is the end of the season and a is age. The decision variables (x_i) that have been used in the model are:

1. planting density
2. timing of the first fruiting branch
3. rate of fruit point production
4. age of transition for the growth rates of single fruit
5. growth rate of young fruit
6. growth rate of older fruit.

Other decision variables might be the leaf to stem ratio during both the vegetative and fruit maturation phases. Powell's method (Powell, 1964) was used in this work so that much of the complexity could be retained. Other optimization procedures are currently being investigated for use on this problem. The value of this type of analysis to plant breeding are obvious, as the x_i are all characters which can be easily observed and probably be selected. Preliminary results indicate that achieving a global optimum is difficult, but that several local optima of approximately the same Y are found. That is, plants can be engineered in several ways to produce the same crop.

If we now wish to use the program to minimize insect damage, we could incorporate the insect model into the optimal (or observed) plant program, and use the same objective function. The decision variables in this case might be quantity and timing of pesticides, planting date for the crop, plant density, or some attribute of plant growth and development.

CONCLUSIONS

The methods of "systems analysis" are useful tools for assembling, analyzing and evaluating data from complex problems (e.g., cotton ecosystem studies). These methods have enabled us to simplify the problem, distill out commonalities between trophic levels, and gain additional insights into the system which were previously unavailable.

While the methods of "systems analysis" are extremely useful, they will always be limited by the quality of the biological data used, the competence of the analyst and the creativity of the scientist. Because of the complexity we must deal with, close cooperation between scientists in impinging disciplines is essential, and more, not less, data of the right kind will be required. This interdisciplinary exchange has been most useful in the development of crop production-crop protection strategies in cotton.

REFERENCES

Andrewartha, H.G. and L.C. Birch (1954), *The Distribution and Abundance of Animals*, Univ. of Chicago Press, Chicago, Illinois, USA.

Auslander, D.M., G.F. Oster and C.B. Huffaker (1974), Dynamics of interacting populations, *J. Franklin Inst. 297*, 5.

Barr, R.O. et al. (1973), Ecologically and economically compatible pest control, in *Insects: Studies in Population Management*, P.W. Geier, L.R. Clark, D.J. Anderson, H.A. Nix, eds., Ecol. Soc. Aust. (memoirs) 1, Canberra, 241-264.

Campbell, A., B.D. Frazer, N. Gilbert, A.P. Gutierrez and M.P. Mackauer (1974), Temperature requirements of some aphids and their parasites, *J. Appl. Ecol.*, 11, 431-438.

De Wit, C.T., R. Brouwer and F.W.T. Penning de Vries (1970), The simulation of photosynthetic systems, in *Prediction and Measurements of Photosynthetic Productivity*, PUDOC, Wageningen, The Netherlands, 47-78.

Duncan, W.G., D.N. Baker and J.D. Hesketh (1971), Simulation of growth and yield in cotton III, a computer analysis of the nutritional theory, *Beltwide Cotton Prod. Res. Conf. Proc.*, 78.

Ehler, L.E., K.G. Eveleens and R. Van Den Bosch (1974), An evaluation of some natural enemies of cabbage looper in cotton in California, *Environ. Ent.*, 2, 1009-15.

Evans, L.T., ed (1975), *Crop Physiology, Some Case Histories*, Cambridge Univ. Press.

Eveleens, K.G., R. Van Den Bosch and L.E. Ehler (1974), Secondary outbreak induction of beet armyworm by experimental insecticide applications in cotton in California, *Environ. Ent.*, 2, 497-503.

Fick, G.W. and R.S. Loomis (1974), Simulation of sugar beet ecology from physiological information in *Crop Physiology, Some Case Histories*, L.T. Evans, ed., Cambridge Univ. Press.

Frank, P.W. (1960), Prediction of population growth form in Daphnia pulex cultures, *Am. Nat.*, 94, 357-372.

Frazer, B.D. and N.E. Gilbert (1976), Coccinellids and aphids: A qualitative study of the impact of adult lady birds (Coleoptera: Coccinellidae) preying on field populations of pea aphids (Homoptera: Aphididae), *J. Ent. Soc. B.C.*, 73.

Gilbert, N.E., A.P. Gutierrez, B.D. Fraser and R.E. Jones (1976), *Ecological Relationships*, Freeman and Co., London.

Griffith, K.J. and C.S. Holling (1969), A competitive submodel for parasites and predators, *Can. Ent.*, 101, 785-818.

Gutierrez, A.P., G.D. Butler, Jr., Y. Wang and D. Westphal (1977), A model for pink bollworm in Arizona and California, *Canad. Ent.* (in press).

Gutierrez, A.P., J.B. Christensen, C.M. Merritt, W.B. Loew, C.G. Summers and W.R. Cothran (1976), Alfalfa and the Egyptian alfalfa weevil, *Canad. Ent.* (in press).

Gutierrez, A.P., L.A. Falcon, W.B. Loew, P.A. Leipzig and R. Van Den Bosch (1975), An analysis of cotton production in California: A model for Acala cotton and the effects of defoliators on its yields, *Environ. Ent.*, 4, 125-136.

Gutierrez, A.P., D.E. Havenstein, H.A. Nix and P.A. Moore (1974), The ecology of Aphis craccivora Koch and subterranean clover stunt virus in Southeastern Australia II, A model of cowpea aphid populations in temperate pastures, *J. Appl. Ecol.*, 11, 1-20.

Gutierrez, A.P., T.F. Leigh, Y. Wang and R. Cave (1977), An analysis of cotton production in California, Lygus hesperus injury - an evaluation, *Canad. Ent.* (in press).

Harper, J.L. and J. White (1974), The demography of plants, *Ann. Review of Ecology and Systematics*, 5, 419-463.

Hesketh, J.D., D.N. Baker and W.G. Duncan (1972), II. Simulation of growth and yield in cotton - environmental control of morphogenesis, *Crop. Sci.*, 12, 436-439.

Hesketh, J.D., D.N. Baker and W.G. Duncan (1971), Simulation of growth and yield in cotton: respiration and the carbon balance, *Crop. Sci.*, 11, 394-398.

Holling, C.S. (1959), The components of predation as revealed by a study of small-mammal predation of the European pine sawfly, *Canad. Ent.*, 91, 292-320.

Holling, C.S. (1966), The functional response of invertebrate predators to prey density, *Mem. Ent. Soc. Can*, 48, 3-86.

Hughes, R.D. (1963), Population dynamics of the cabbage aphid, Brevicoryne brassicae (L.), *J. Animal Ecol.*, 32, 393-426.

Hughes, R.D. and N. Gilbert (1968), A model of an aphid population - general statements, *J. Anim. Ecol.*, 37, 553-563.

Jones, J.W., H.D. Bowen, J.R. Bradley and R.E. Stinner (1974), Modelling interactions of boll weevils on cotton crops by analysis of behavioral patterns, *Beltwide Cotton Prod. Res. Conf. Proc.*, 102-103.

Leslie, P.H. (1945), On the use of matrices in certain population mathematics, *Biometrika,* 33, 183-212.

Lotka, A.J. (1925), *Elements of Physical Biology,* Williams and Wilkins, Baltimore.

McKinion, J.M., J.W. Jones and J.D. Hesketh (1974), Analysis of SIMCOT: Photosynthesis and growth, in *Beltwide Cotton Prod. Res. Conf. Proc.*, 117-124.

McKinion, J.M., J.W. Jones and J.D. Hesketh (1975), A system of growth equations for the continuous simulation of plant growth, *Am. Soc. Agric. Eng. paper,* 18, 975-979 and 984.

Miles, G.E., R.J. Bula, D.A. Holt, M.M. Schreiber and R.M. Peart (1973), Simulation of alfalfa growth, *Am. Soc. Agric. Eng.*, paper 73, 4547.

National Academy of Sciences (1976), Committee on climate and weather fluctuations and agricultural production (to be released soon).

Nicholson, A.J. and V.A. Bailey (1935), The balance of animal populations, *Proc. Zool. Soc.*, Pt. 1, London, 551-598.

Powell, M.J.D. (1964), An efficient method for finding the minimum of a function of several variables without calculating derivatives, *The Computer Journal,* 7, 155-162.

Regev, U., A.P. Gutierrez and G. Feder (1976), Pests as a common property resource: A case study of alfalfa weevil control, *Amer. J. Agric. Econ.*, 58, 187-197.

Rogers, D.J. and M.P. Hassell (1974), General models for insect parasite and predator searching behavior: interference, *J. Anim. Ecol.*, 43, 239-253.

Royama, T. (1970), Factors governing the hunting behavior and selection of food by the great tit. Parsus major L., *J. Anim. Ecol.*, 39, 619-668.

Sevacherian, V. and V.M. Stern (1972), Spatial distribution patterns of Lygus bug in California cotton fields, *Environ. Entomol.*, 1, 695-703.

Sharpe, P.J.H. and D.W. DeMichele (1977), Reaction kenetics of poikilotherm development, *J. Theor. Biol.* (in press).

Sinko, J.W. and W. Streifer (1967), A new model for age-size structure of a population, *Ecology*, 48, 910-918.

Slobodkin, L.B. (1954), Population dynamics in Daphnia obtusa Kurz, *Ecol. Monogr.*, 24, 69-89.

Stapleton, H.N., D.R. Butson, F.L. Watson, D.L. Nolting and D.N. Baker (1973), *Cotton: A Computer Simulation of Cotton Growth*, Tech. Bulletin Ariz. Agric. Expt. Sta.

Stinner, R.E., A.P. Gutierrez and G.D. Butler, Jr. (1974), An algorithm for temperature-dependent growth rate simulation, *Can. Ent.*, 106, 519-524.

Stinner, R.E., R.L. Rabb and J.R. Bradley (1974), Population dynamics of Heliothis zea (Boddie) and H. virescens (F) in North Carolina: a simulation model, *Environ. Entomol.*, 3, 163-168.

Streifer, W. (1975), Realistic models in population ecology, *Adv. Theoretical Ecology*, 199-266.

Thompson, W.R. (1924), La theorie matematique de laction des parasites entomophages et le factem du hasard, *Am. Fac. Sci. Marsellie*, 2, 69-89.

Volterra, W. (1931), Lecous sur la theorie matematique de la lutle pour la Vic., Gauthier-Villars, Paris.

Von Foerster, H. (1959), Some remarks on changing populations, in *The Kinetics of Cellular Proliferation*, F. Stohlman, Jr., ed., Grune and Stralton, New York.

Wang, Y., A.P. Gutierrez and G. Oster (1977), A general model for plant growth and development: coupling cotton-herbivore interactions, *Can. Ent.* (in press).

Watt, K.E.F. (1966), The nature of systems analysis, in *Systems Analysis in Ecology*, K.E.F. Watt, ed., Academic Press, New York, 1-14.

ECONOMIC CONFLICTS IN PLANT PROTECTION: THE PROBLEMS OF PESTICIDE RESISTANCE; THEORY AND APPLICATION TO THE EGYPTIAN ALFALFA WEEVIL[1]

Uri Regev[2], Haim Shalit[3] and A. P. Gutierrez[4]

INTRODUCTION

The use of chemical pesticides in agriculture against various pests (insects, mites, weeds, fungi, etc.) has resulted in the development of resistance to the chemicals by the pest, and has created additional severe problems in pest management. The economic problem of optimal allocation of pesticides to control specific insect pests has been studied by Regev et al. (1976) and Shoemaker (1976), while the problems created by pesticide resistance have been examined only to a very limited extent (Taylor et al., 1975; and Hueth and Regev, 1974). The importance of the problem is so great and general that a popularized article recently published in Time magazine (July 12, 1976), reported, "Because of over exposure [to insecticides], insects are becoming more immune . . . . . ." Experience has shown that as resistance develops, farmers use ever greater amounts of increasingly expensive pesticides to control pests. This situation leads to a vicious circle, wherein all of the spin-offs are negative (e.g., increased costs, pesticide pollution, secondary pest outbreaks, wild life destruction, etc.). The magnitude of the externalities caused by pesticide use are difficult to estimate, and the additional costs incurred as a result of pesticide resistance,

---

[1]This work was supported, in part, by the National Science Foundation and Environmental Protection Agency through a grant (NSF-GB-347181, BMS-74-04223) made to the University of California. The findings, opinions, and recommendations expressed herein are those of the authors and not necessarily those of the University of California, the National Science Foundation, or the Environmental Protection Agency.

[2]Department of Economics, Ben Gurion University, Beersheva, Israel.

[3]Department of Agricultural and Resource Economics, University of California, Berkeley, U.S.A.

[4]Division of Biological Control, University of California, Berkeley, U.S.A.

are unknown. Environmental damage has rapidly accrued, but the damage is not irreversible; susceptability in the pest population can be considered a resource which is not renewable once it has been depleted. Entomologists and other field scientists have developed new control methods (e.g., plant breeding, biological controls, cultural practices, etc.) to cope with specific pests and, by default, with the development of resistance in the pest population.

This paper attempts to develop a general theoretical model for the optimal allocation of pesticides given that resistance has developed in the pest population. The work emphasizes the need for regional organization or governmental intervention, (i) to reduce the negative externalities created by individual grower pesticide use, (ii) to extend the effectiveness of the pesticide until a new alternate technology can be developed and, (iii) to estimate the developmental investments that should be made at any time t in alternate technologies. This general model is based upon an empirical study (Regev et al., 1976, and Gutierrez, Regev and Shalit, in progress), and is constrained to a single pest—single crop framework, dealing with the case of simple, single gene resistance to a specific pesticide. Many of the known cases of resistance to pesticides are of this kind (Georghiou, 1972; Brown, 1967). The extension of the problem to a multigene-multitoxicant case will not necessarily add to the results obtained in this paper.

The conclusions drawn from the model are two-fold; (i) by assuming the existence of alternative pest control techniques, the optimal path of current pesticide practices is found until the economy switches to one of the alternative technologies, (ii) given the present insecticide technology, the value of investment in new pest control methods is determined. An empirical study, dealing with alfalfa and the Egyptian alfalfa weevil, is discussed to accommodate the theoretical analysis.

THE MODEL

This section presents a simplified, continuous time model of the economic problems associated with the development of pesticide resistance in pest populations. The pest management system considered here is a region closed to external pest migration. It is presumed that a central decision body, which has perfect information on the pest-plant control system, maximizes the present value of profits for a given crop. All production factors except pest control are assumed to be at their optimal levels and are not influenced by pesticide application (the control variable).

w(t) is defined as the level of resistance at time t, measured as the proportion of pest population immune to a particular pesticide or group of pesticides. Once resistance appears, it changes over time as a result of pesticide application, x(t), according to the following rule:

$$\frac{dw}{dt} \equiv \dot{w}_t = g(x_t, w_t) \tag{1}$$

where†

$$g_x \geqq 0, \quad g(o, w) = 0, \quad \lim_{w \to 1} g(x, w) = 0$$
$$\text{for all } x \geqq 0, \quad 0 < w < 1 \quad . \tag{2}$$

The function, g(·), relates resistance development from some level at t to subsequent pesticide applications and assumes w to be nondecreasing in x. The results of this optimization problem, concerning the interaction of pesticide application and the development of pesticide resistance in a population considered in the model, are not restricted to a single pest or pesticide. The history of pest management has shown that, whatever pesticide technologies are used, decreasing pest susceptibility is an expected outcome (i.e., the development of resistance). Very little is known about the explicit form of the function, g(·). However, from empirical studies, an indirect estimate of this relationship can be obtained (as in the Empirical Study section). Furthermore, the assumption, g(o, w) = 0, together with $g_x \geqq 0$ and $x \geqq 0$, implies that w(t) is a nondecreasing function, i.e., resistance does not decrease over time once it has arisen.

Entomologists have observed that, in some cases, resistant pests are less able to survive in the absence of pesticides than are the nonresistant ones (see Georghiou, 1972). Thus, in some cases, the resistance level might decrease in the absence of pesticides. However, the rate of decrease is slow relative to its increase once pesticide use restarts and, in the long-run, it seems likely that one can disregard the short-run decrease of resistance and consider the biological process irreversible.

Let $\pi(x_t, w_t)$ be the profit function generated through the use of pesticides (disregarding other inputs for simplicity). One assumes the following properties for π(·):

$$\pi_w \leqq 0, \quad \pi_{xx} < 0, \quad \pi(o, w) = \text{constant}. \tag{3}$$

†Subscripted variables (but not t) denote partial derivatives, and t is omitted for notational convenience whenever possible with no ambiguity.

The negative effect of increasing resistance on profits results from the increased quantities of insecticides needed to reduce the pest population.

The current pest control technology (x) is maintained until some future time T when a readily available alternative technology will be used. This technology (α) could be either an alternative set of chemical pesticides, biological control, any mixture of the two, or pest-resistant plants. The alternative technology (α) yields at time T a net return of

$$S(\alpha, w_T) = \max \int_T^{\infty} s[\alpha, y_t(w_T)]\, e^{-rt}\, dt \quad ; \qquad S_\alpha \geq 0,\ W_{w_T} \leq 0 \tag{4}$$

where

$y_t(w_T)$ = control of the α technology

$s(\cdot)$ = net benefits stream

and

r = fixed discount rate.

Since this model focuses mainly on the current technology and the switching date, the form and behavior of $S(\cdot)$ are not investigated. $S(\alpha, w_T)$ is taken as the future profits discounted to time T. However, if the alternative process does not, in any context, increase pesticide resistance in the population, it is the "backstop technology" (Nordhaus, 1975) with respect to pesticides. It is further assumed that S is differentiable in α, and technologies are ranked so that $S_\alpha \geq 0$. The effect of pesticide resistance to current technology on the future α technology is nonpositive, $S_w \leq 0$.

Optimal Solution

The economic problem is now defined as the choice of quantities of pesticide (x_t) to be applied and of the date of switching to the new technology which maximizes

$$\phi(T) = \int_0^T \pi(x_t, w_t)\, e^{-rt}\, dt + e^{-rT} S(\alpha, w_T) \tag{5}$$

subject to equation (1) and $x_t \geqq 0$. Applying the Maximum Principle, there exists a continuous function, $\lambda(t)$, and a function, $\mu_t \geqq 0$, which satisfy the necessary conditions for a maximum. For a fixed endpoint, t = T, these conditions are

$$\pi_x + \lambda_t g_x + \mu_t = 0; \qquad \mu_t \geqq 0, \qquad \mu_t x_t = 0 \tag{6a}$$

$$\dot{\lambda}_t = (r - g_w)\, \lambda_t - \pi_w \tag{6b}$$

$$\dot{w} = g(x, w); \qquad w_o \text{ is given} \tag{6c}$$

$$\lambda_T = \frac{\partial S(\cdot)}{\partial s_T} \text{ (transversality condition).} \tag{6d}$$

Equation (6a) is the usual marginal condition of profit maximization. In the economically relevant range, $\pi_x \geqq 0$, and since $g_x \geqq 0$, λ_t is nonpositive. This is the marginal cost of resistance. Furthermore, $\lambda_t g_x$ represents the marginal future benefits foregone by pesticide application at t. This is the user cost of pesticides (Scott, 1967, p. 34). The role of λ_t as an economic price is further clarified by equation (6d); namely, at the switching time T, the marginal cost of resistance equals the reduction in future benefits if the α technology is not used. Equation (6b) determines the behavior of the time path of the resistance cost, λ_t.

So far, the switching date has been arbitrarily fixed. However, assuming $\phi(T)$ differentiable and concave in T, the optimal switching date T* is found by $d\phi^*(T)/dT = 0$, where $\phi^*(T)$ is optimal with respect to x_t [i.e., satisfies conditions (6)].

$$\frac{d\phi}{dT} = \int_0^T \left(\pi_x \frac{\partial x_t}{\partial T} + \pi_w \frac{\partial w_t}{\partial T}\right) e^{-rt}\, dt + e^{-rT} \left[\pi(x_T, w_T) - rS(\alpha, w_T) + S_w \frac{\partial w_T}{\partial T}\right] . \tag{7}$$

Note that

$$\frac{\partial w_t}{\partial T} = \int_0^t g_x \frac{\partial x_\tau}{\partial T} + g_w \frac{\partial w_\tau}{\partial T} \; d\tau$$

and

$$\frac{\partial w_T}{\partial T} = \dot{w}_T = g(x_T, w_T) \quad .$$

Insert conditions (6) and (7) to obtain

$$\frac{d\phi^*(T)}{dT} = e^{-rT} \; [\pi^*(x_T, w_T) - rS(\alpha, w_T) + \lambda \dot{w}_T]$$

$$- \int_0^T \lambda_t \left(g_x \frac{\partial x_t}{\partial T} + g_w \frac{\partial w_t}{\partial T} \right) e^{-rt} \, dt \tag{8}$$

$$- \int_0^T (\dot{\lambda} - r\lambda) \, e^{-rt} \frac{\partial w_t}{\partial T} \, dt - \int_0^T e^{-rt} \mu_t \frac{\partial x_t}{\partial T} \, dt \quad ,$$

where π^* indicates the optimal value of π. The last term in (8) vanishes since $\mu_t \partial x_t / \partial T = 0$.† Integrating by parts, it follows from $d\phi^*(T)/dT = 0$ and $d^2\phi/dT^2 < 0$ that

$$\pi^*(x_{T*}, w_{T*}) = rS(\alpha, w_{T*}) \, , \; \frac{d\pi^*}{dT} - rS_w \frac{\partial w}{\partial T} \leq 0 \quad . \tag{9}$$

By (9), T* is optimal if the immediate gain (π^*) of postponement equals the instantaneous future loss $rS(\cdot)$.

---

†Since $\mu_t x_t = 0$ implies either $x_t = 0$ or $\mu_t = 0$ for all t, then

$$\frac{\partial \mu_t x_t}{\partial t} = \mu_t \frac{\partial x_t}{\partial t} + x_t \frac{\partial \mu_t}{\partial t} = 0 \text{ for all t and implies}$$

$$\mu_t \frac{\partial x_t}{\partial t} = -x_t \frac{\partial \mu_t}{\partial t} = 0 \quad .$$

Equations (6) and (9) are the necessary conditions for the optimal pest control policy. The uniqueness of the solution (if it exists) is guaranteed by the concavity of the optimal Hamiltonian with respect to w_t.

From equation (9) and by $\dot{w} \geqq 0$ and $\pi_w \leqq 0$, it follows that:

Proposition 1. The new technology (α) will never be adopted (i.e., $T^* \rightarrow \infty$) if $\pi(x, w) > rS(\alpha, w)$ for all x, w, and T. On the other hand, if $\pi(x_o, w_o) \leqq rS(\alpha, w_o)$ for all $x_o \geqq 0$, then $T^* = 0$.

Thus, one can a priori determine the minimal value of the α technology ($\overline{S}$) which should be adopted immediately. Furthermore, given the current practices, the policy-maker can find and exclude all the nonprofitable alternatives ($S \leqq \underline{S}$). Figure 1 presents the optimal path of π^* together with lower and upper bounds ($\underline{S}$, $\overline{S}$) as expressed by Proposition 1.

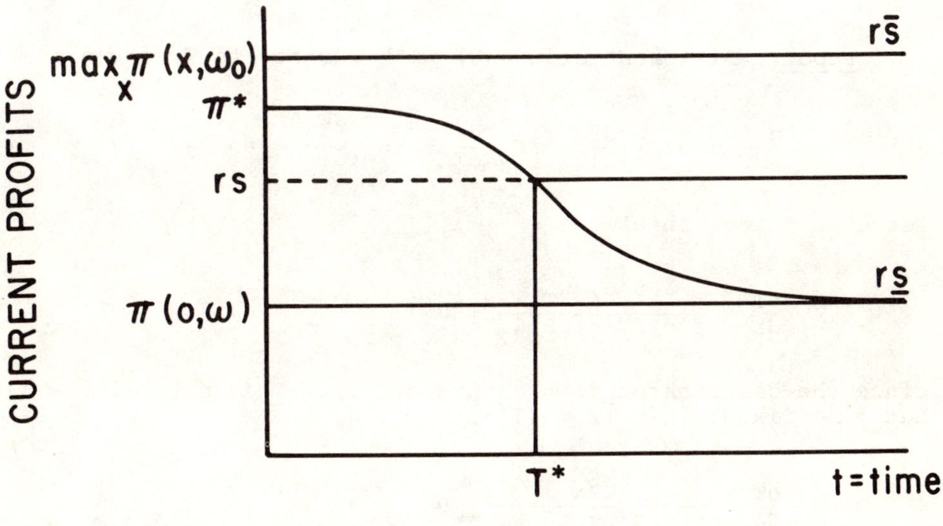

Figure 1. The optimal path of profit (π^*) overtime. (See text for discussion of all variables).

Analyzing the sensitivity of the optimal switching date T* with respect to some parameters of the model, it follows that:

Proposition 2. If an optimal program exists, then the optimal switching date is advanced (a) by an increase in the interest rate:

$$\frac{dT^*}{dr} \leqq 0 \quad ; \tag{10a}$$

(b) by technological improvement:

$$\frac{dT^*}{d\alpha} \leqq 0 \quad ; \tag{10b}$$

and (c) by a rise in the initial level of resistance:

$$\frac{dT^*}{dw_o} \leqq 0 \quad . \tag{10c}$$

Proof. Differentiating (9) with respect to T, one gets

$$\frac{d\pi^*}{dT} - S(\cdot)\,\frac{dr}{dT} - rS_\alpha\,\frac{dx}{dT} - rS_w\,\frac{dw_T}{dT} = 0 \quad . \tag{11}$$

Let α be fixed; then

$$\frac{dT}{dr} = \frac{S(\alpha,\ w_T)}{d\pi^*/dT - rS_w\dot{w}_T} \leqq 0 \tag{12}$$

since the demoninator is negative by (9). This establishes (10a). Let r be fixed; then from (11),

$$\frac{dT}{d\alpha} = \frac{rS_\alpha}{d\pi^*/dT - rS_w\dot{w}_T} \leqq 0 \tag{13}$$

which establishes (10b). To prove (10c), one only needs to invoke Bellman's (1957) principle of optimality. Since the system is autonomous, starting with any $\tilde{w} > w_o$ yields the same optimal solution which is obtained from the original problem from $\overline{t}$ to T*, where $\overline{t} = \{t\ (0,|T^*)\ w^*(t) = \tilde{w}\}$. Thus,

$$T^*(\tilde{w}) = T^*(w_o) - t(w_t) \leqq T^*(w_o) \text{ for all } \tilde{w} \geqq w_o. \qquad \text{Q.E.D.}$$

Furthermore, if pesticide resistance is nondecreasing, even in the absence of pesticide application, it is optimal to spray continuously until the switching date.

Proposition 3. For all $t < T$, $x_t = 0$ is never optimal.

Proof: The proposition will be proved by showing that if, for any t', $x^*_t = 0$, then $t' \geqq T^*$. Suppose the contrary, i.e., $x^*_{t'} = 0$ and $t' < T^*$; then

$$\pi(0,\ w_{t'}) > rS(\alpha,\ w_{t'}) \quad ; \tag{14}$$

but $rS(\alpha,\ w_{t'}) \geqq rS(\alpha,\ w_T)$ since $S_w \leqq 0$ and $w_{T*} \geqq w_{t'}$.

However, $\pi(0,\ w_{t'}) = \pi(0,\ w_{T*})$ by (3), and $\pi(0,\ w_{T*}) \leqq \pi(x_{T*},\ w_{T*})$ since $\pi_x \geqq 0$ in the relevant range. This establishes by equation (9):

$$\pi(0,\ w_{t'}) \leqq rS(\alpha,\ w_{T*}) \quad . \tag{15}$$

But equation (14) contradicts (15) Q.E.D.

Proposition 3 suppresses the possibility of pesticide rotation in the same crop since, under the assumption $g(0,\ w) = 0$, resistance to one insecticide does not decrease when the second is sprayed. However, if $g(0,\ w) < 0$, proposition 3 does not necessarily hold. By insecticide rotation or by small pesticide applications, a steady state in the resistance level may be feasible. If a steady state is reached, new alternatives will not be adopted unless technological changes improve their net revenues. However, the current pest control practices do not indicate that it is feasible to keep a steady state resistance level.

A Comparison with a Competitive Solution

A single decision unit faces basically the same problem of pest control as described above. However, since the pests are migrating, resistance develops in respect to the total pesticides applied in the region. Therefore, the effect of the individual farmer on resistance development is negligible and must be considered to be beyond his control. Because the decision maker does not affect the general level of resistance, he disregards it and seeks to maximize his immediate net benefits. His decision, however, depends on the resistance level which he presumably observes. The competitive decision maker then applies pesticides at the level of $\hat{x}_t(w_t)$ to maximize

$\pi_x(x_t, w_t)$. Assuming, as before, that a new technology (α) is available at any time, he will switch to it whenever $\pi(\hat{x}_t, w_t) \leqq rS(\alpha, w_t)$. This solution requires that $\pi_x(\hat{x}_t, w_t)=0$ for all t. His switching date, T, is such that $\pi(\hat{x}_t, w_t) = rS(\alpha, w_t)$.

Comparing (6a) with the competitive solution, it is clear that, for any given level of pesticide resistance, the competitive solution implies a higher level of pesticide application than that desired by society since $\pi_{xx} \leqq 0$ and $\lambda_t g_x \leqq 0$. This is a known result for negative externalities. Note also that, for any given level of pest resistance, the competitive solution ($\hat{x}_t$) yields an immediate net gain higher than the optimal one. However, since $\hat{x}_t(w) \geqq x^*_t(w)$ and $g_x \geqq 0$, pest resistance levels in the competitive situation are higher than in the centralized solution, i.e., $\hat{w}_t \geqq w^*_t$ for all t. Considering the resistance levels at the two switching dates, it is shown that:

Proposition 4. $w^*_{T^*} \leqq \hat{w}_{\hat{T}}$, and equality holds if, and only if, $S_w = 0$ (or $g_x = 0$).

Proof. Recall that, for any given w, $x^*(w) \leqq \hat{x}(w)$ and $w^*_t \leqq w_t$, for all $t \leqq \min(\hat{T}, T^*)$.

If $T^* < \hat{T}$, then $w^*_{T^*} \leqq \hat{w}_{T^*} \leqq \hat{w}_{\hat{T}}$ since $\dot{w} \geqq 0$.

If $T^* \geqq \hat{T}$, then for $\hat{T} \leqq t \leqq T^*$, by (9):

$$rS(\alpha, w^*_t) \leqq \pi^*(w^*_t) \leqq \max_x \pi(x, w^*_t) \quad . \tag{16}$$

At $t = \hat{T}$, $\hat{w}_t \geqq w^*_t$. Let $\bar{T}$ be an arbitrary point greater than T*, and let t be a sequence of points tending to $\bar{T}$. Since w(t) is continuous and non decreasing, there exists a point t' such that $\hat{w}_{\hat{T}} = w^*_{t'}$. Then

$$rS(\alpha, w^*_{t'}) = rS(\alpha, \hat{w}_{\hat{T}}) = \hat{\pi}(\hat{w}_{\hat{T}}) = \max_x (\hat{x}, w_{\hat{T}}) = \max_x (x, w^*_{t'}) \tag{17}$$

or

$$rS(\alpha, w^*_{t'}) = \max_x \pi(x, w^*_{t'}) \quad .$$

Then, by equation (16), $rS(\alpha, w_{t'}) = \pi^*(w^*_{t'})$, and by equation (9), t' = T*. For all the points, $t'' > T^*, \hat{w}_{\hat{T}} < w^*_{t''}$. However, this cannot hold since $S(\cdot)$ and $\pi^*(\cdot)$ are continuously non increasing with time. To prove the second part, note that, if

$S_w < 0$ (and $g_x > 0$), then $\lambda_T < 0$ and $\lambda_t g_x < 0$ for all $t \leqq T$. Therefore, by (6), the rightmost inequality of (16) must be a strict inequality, and $\pi^*(w) < \max_x \pi(x, w) = \hat{\pi}(w)$ for all w.

Therefore, (17) cannot hold and $w^*_{T^*} < \hat{w}_{\hat{T}}$. Since $S_w \leqq 0$ and $w^*_{T^*} \leqq \hat{w}_{\hat{T}}$, this is equivalent to $w^{**}_T = \hat{w}_{\hat{T}}$ implying that $S_w = 0$. On the other hand, if $S_w = 0$, then by (6d) $\lambda_T = 0$ and by (6a) and (9) at T*:

$$\max_x \pi(x, w_{T^*}) = \pi^*(w^*_{T^*}) = rS(\alpha, w^*_{T^*}) \quad . \tag{18}$$

By the uniqueness of $\max_x \pi(x, w)$, the rightmost equality of (18) implies by (9) that $w^*_{T^*} = \hat{w}_{\hat{T}}$. Q.E.D.

Since the private solution implies a resistance level not lower than the one of the optimal solution at their respective switching dates, one cannot generally determine which switching date is earlier. Nevertheless, if $S_w = 0$, the two-terminal resistance levels are equal in both solutions by Proposition 4; and since $w^*_t \leqq \hat{w}_t$ for all t (and $\dot{w}_t \geqq 0$), one can conclude:

Proposition 5. $S_w = 0$ implies $\hat{T} \geqq T^*$.

Conditions of proposition 5 are likely to prevail in the case of biological control which is not affected by the immunity level toward chemical pesticides.

AN EMPIRICAL STUDY

The case study involves the problem of Egyptian alfalfa weevil [*Hypera brunneipennis* - EAW] which is a major alfalfa pest in California. In the late 1950's, this pest invaded the Southwestern parts of the United States causing an estimated annual damage of over $18 million in California alone. Though resistance to pesticides has not yet developed in this species in California, the threat is imminent as evidenced by the evolution of hexachlor resistance in a closely related species in the USSR (Mardzhangan et al., 1969).

The Biological Context

A detailed description of the biological systems underlying the model is presented in an article by Gutierrez, Regev and Shalit (in preparation). The mathematical formulation is based on a similar model (Regev, Gutierrez, and Feder, 1976), and uses field and laboratory data from a study by Gutierrez et al. (1976). A brief summary of these papers is presented here and interested readers should refer to them for further details.

The biology of the pest and the alfalfa crop (i.e., hay) is briefly described. The adult weevils migrate into the alfalfa fields from aestivation (summer dormant period) in the autumn and early winter months and oviposit their eggs. The eggs hatch to produce larvae which feed on the leaf tissues of the alfalfa.

The larvae which survive the winter and spring frost can cause damage. The larvae develop to the pupal stage and later in the spring become young new adults which leave the fields and remain dormant during the summer, returning the following autumn to begin the cycle anew.

Farmers commonly apply chemicals (e.g., Furadan, R) when feeding damage is observed, to control adult and larval populations. The effect of the pesticide is formulated by a dosage response function which depends upon stage (of adult or larva) and upon the level of pesticide resistance (w) in the population. In the model, resistance is controlled by a single dominant gene, and susceptibility in homozygous and heterozygous individuals differ (see Gutierrez, Regev, Shalit, in preparation). Resistance levels in the population are measured as the relative frequency of the resistance gene in the population. Changes in gene frequency are determined by pesticide-induced, genetic selection, following the Hardy-Weinberg law. During the aestivation period (summer), adults from many fields mix together so that the frequency of the resistant gene depends on the total amount of pesticide applied in the whole region, and not upon the application made by single farmers.

The Optimization Procedure

The seasonality of the pest and plant life-cycle and their interaction determine, to a large extent, the type of optimization procedure. The solution to the problem requires two steps: (i) the intraseasonal solution is obtained through a formulation similar to that of Regev, Gutierrez and Feder (1976), but incorporates pesticide resistance, after which (ii) the one-season solutions are used in the interseasonal problem to provide the optimal solution for the entire model. The intraseasonal or single season problem is summarized as:

$$\begin{aligned} &\max \pi_t(\underline{x}_t, w_t) \\ &\text{subject to} \quad h_t^j(\underline{x}_t, w_t) = 0 \quad j = 1 \ldots J \end{aligned} \tag{19}$$

where

π_t is the profit function at time t,

$\underline{x}_t$ is the amount of pesticide applications,

w_t is the frequency of the resistance gene in the adult weevils, and

j denotes within season time and t between seasons.

The set of difference equations $h_t^j(\cdot) = o$ describe the intraseasonal pest-plant dynamics, and the intraseasonal changes in w_t, both of which are affected by the amount and timing of pesticide applications. At the end of the season, the frequency of the resistance gene in the new summer adults leaving the fields is given by w_{t+1}. Therefore, one can formulate the interseasonal development of pesticide resistance as:

$$w_{t+1} = w_t + g(\underline{x}_t, w_t) \tag{20}$$

which is the discrete form of equation (1). Because of computer core limitations, the procedure was to solve the system (19) subject to equation (20) for different constrained values of w_t and w_{t+1} as follows:

$$\phi(w_t, w_{t+1}) = \max_{\underline{x}_t} \pi(\underline{x}_t, w_t) \tag{21}$$

$$\text{subject to: } h_t^j(\underline{x}_t, w_t) = 0$$

$$w_{t+1} = w_t + g(\underline{x}_t, w_t) \quad ,$$

$$\text{for some given } w_t \text{ and } w_{t+1} \quad .$$

The multiseasonal problem is then:

$$\psi(t) = \max_{w_t} \sum_{t=o}^{T-1} {}^{t} \quad \phi(w_t, w_{t+1}) + \beta^t S(\alpha, w_t) \tag{22}$$

for some given w_o, $o<w_t < 1$

where

$\beta^t = 1/(1+r)^t$ j , T is the planning horizon and $S(\cdot)$ is defined by equation (4).

The values of $\phi(w_t, w_{t+1})$ are obtained for all relevant values of w_t and w_{t+1} by using the non-linear programming code (GRG), while the solution to (22) is obtained through a discrete dynamic programming technique as follows:

Let n = T-t. Define

$$f_n(w_t) = \max\ [\phi(w_t, w_{t+1}) + \beta f_{n-1}(w_{t+1})] \tag{23}$$

$$f_o(w_t) = S(\alpha, w_t)\ .$$

By a recursive procedure, a solution to (23) is found which is identical to the multiseasonal problem (22), and is the optimal solution. The optimal switching date T* is determined by choosing N in (23) which maximizes $f_n(w_o)$. Then T* has to satisfy the following conditions:

$$\pi^*(w_{T^*-1})\ (1+r) > rS(\alpha, w_{T^*-1}) \tag{24}$$

$$\pi^*(w_{T^*})\quad (1+r) \leqq rS(\alpha, w_{T^*})\ .$$

(24) is similar to condition (9) in the continuous time model. However, the term (1+r) is needed since the new technology is employed one season after the current practices are ceased.

RESULTS

Using the maximization procedure described above, Table 1 presents the solution to the single season problem (21) for different values of w_t and w_{t+1}. The profit in dollars per acre appears in the third column, the total amount of pesticide (in ounces per acre) applied during the season is indicated in the fourth column, and the last column shows the shadow price $(-\lambda)$ of the constraint (20), i.e., the marginal cost of pesticide resistance.

From Table 1, the upper and lower bounds of the short run profits are found to be \$116.87 and 73.0. The upper bound is obtained when $w_t = w_o = .01$ and w_{t+1} is unconstrained, while the lower bound estimates the profits when no pesticides are applied, regardless of resistance level. These bounds determine the values of any new technology as expressed in proposition 1 and plotted in Figure 1.

For a real interest rate of 5% (r = .05) the minimal value $(\bar{S})$ of an alternative technology to be immediately adopted (i.e., T* = 0) is \$2456. The upper bound for non-profitable alternatives is $\underline{S}$ = \$1533, i.e., for any alternative for which $S \leq \underline{S}$ T* tends to infinity and the "no pesticide" alternative is preferred to it.

Table 1. The general results for a single season problem given w_t and w_{t+1}, where π is profit, Σx are the total of pesticides applied within the season and $-\lambda$ is the shadow price of the resistance level w_{t+1}.

w_t	w_{t+1}	π_t	Σx	$-\lambda$
.01	.01	73.0	0	3964.0
.01	.05	116.17	23.24	98.8
.01	.0667	116.87	26.14	0
.05	.05	73.0	0	777.6
.05	.10	96.42	31.3	308.4
.05	.15	106.8	19.5	96.36
.05	.20	112.8	24.31	45.07
.05	.25	114.64	25.6	12.37
.05	.276	114.82	27.0	0
.10	.10	73.0	0	380.09
.10	.15	87.23	7.2	189.13
.10	.20	96.52	16.73	155.17
.10	.25	104.29	23.50	97.98
.10	.35	110.22	24.9	30.92
.10	.473	111.9	28.6	0
.15	.15	73.0	0	213.0
.15	.20	83.1	5.8	163.19
.15	.25	90.1	10.2	100.0
.15	.35	100.18	23.64	83.49
.15	.50	107.46	26.37	21.61
.15	.61	108.55	30.4	0
.20	.20	73.0	0	180.6
.20	.25	80.75	5.1	135.0
.20	.35	90.84	12.9	65.2
.20	.50	100.0	25.0	51.76
.20	.65	101.78	33.27	2.81
.20	.70	104.64	31.60	0
.25	.25	73.0	0	140.42
.25	.35	84.49	7.6	78.71
.25	.50	91.96	16.0	41.6
.25	.65	98.35	26.5	30.55
.25	.768	100.14	31.7	0
.35	.35	73.0	0	93.86
.35	.50	84.25	9.76	44.10
.35	.65	88.75	17.88	15.87
.35	.805	90.01	28.7	0
.50	.50	73.0	0	57.16
.50	.65	80.1	9.76	28.15
.50	.778	82.12	18.02	0
.65	.65	73.0	0	33.29
.65	.80	76.28	11.29	3.10
.65	.811	76.31	12.3	0
.80	.80	73.0	0	6.45
.80	.812	73.8	2.3	0

The question of optimal switching time can be analyzed for a specific type of alternative technology, like biological control or plant breeding. In any case, the value of the new technology does not depend on the final level of resistance (i.e., $S_w = 0$). The choice $S_w = 0$ is dictated by the difficulty of obtaining valid estimates for S_w, and by the propensity of entomologists to use these alternatives to solve resistance problems.

If one assumes first that no alternative technology is available, then the optimal switching time occurs when it is no longer profitable to apply pesticides. This is equivalent to the case when one adopts an available technology with present value $1533. Let's evaluate the optimal path for this hypothesis (i.e., S = 1533, r = .05) by applying method (23) to the results of Table 1. The optimal path is compiled in Table 2 and compared to the path followed by a competitive entrepreneur. Under centralized decision making, it takes six seasons to reach a resistance level of .80, while under competition, the same level is attained in approximately three seasons. Moreover, for each level of resistance, one can observe that generally the competitive grower uses larger amounts of pesticides than the centralized unit. Thus, the stock of pesticide susceptibility is exhausted sooner under competition than under a centralization framework. Propositions 4 and 5 are intuitively proved within the empirical example. Note that the marginal cost of resistance ($-\lambda$) decreases as the resistance level increases over time since less is gained in the future by controlling the resistance level at the present. Moreover, the profit is monotonically decreasing over time.

Let's now assume that an alternative technology is available. By choosing its present value between $\bar{S}$ and $\underline{S}$ one insures that the switching time will be within a finite range. Table 3 provides two optimal paths with two different values for the alternative technology. For S = 1600, the switching time occurs after four seasons, while if S = 2000, it occurs at t = 3, which confirms proposition 2(b). However, within the economically valid range of interest rates, r = .05 to r = .10, the optimal solution is found to be non-sensitive. An interest rate of 20% is needed to advance the switching date, see proposition 2(a). Proposition 2(c) is confirmed, because an increase in w_o from .01 to .05 does not change the optimal path, but does move the switching time forward.

The value of this general model is that it suggests policy for the management of resistance problems based upon estimates of the discounted sums of future profits (i.e., by comparing alternative policies—competitive versus centralized decision making). In the specific case of the Egyptian alfalfa weevil, the difference between the two policies in present value is $9.2 per acre. It would thus appear that government intervention would be economically relevant if the monetary costs of such an intervention are less than the net gains.

Table 2. Comparison of the optimal solution and the competitive solution for $S \leq 1533$ and $r = .05$.

t	0	1	2	3	4	5	6
			The Optimal Solution				
w	.01	.05	.15	.35	50	65	80
$-\lambda$	98.8	96.36	83.49	44.10	28.15	3.10	0
Σx	23.24	19.5	23.64	9.76	9.76	11.29	2.3
π	116.17	106.8	100.18	84.25	80.1	76.28	73.8
			The Competitive Solution				
w		.01	.05	.25	.80		
$-\lambda$		0	0	0	0		
Σx		26.14	27.	31.7	0		
π		116.81	114.82	100.14	73.		

Table 3. Optimal solutions for two levels of benefits from alternative technologies (S), $r = .05$.

	t = 0	1	2	3	4
			S = 1600		
w	.01	.05	.10	.25	.768
π	116.17	96.42	104.29	100.14	Switch to new technology
			S = 2000		
w	.01	.05	.20	.70	
π	116.17	112.8	104.64	Switch to new technology	

POLICY IMPLICATIONS

Since, in the competitive system, farmers apply too much pesticides, government intervention is necessary to reduce them to the optimal quantities. In this framework two classical approaches to internalize the externalities generated by perfect competition can be followed:

1. A Pigouvian tax may be levied per unit of pesticide use. From equation (6a), a tax rate equal to $\lambda_t g_x$ will induce the competitive decision-maker to adopt the optimal pest control policy.

2. Because of the difficulty to evaluate the time path of g_x, it could be preferable to impose restrictions on pesticides that would force the farmer to use socially optimal amounts. However, the procedure involves higher administrative costs of enforcement.

If biological control is the alternative technology, resistance to chemical pesticide would not affect the future benefits ($S_w = 0$). Then, by proposition 5, the switching date for the competitive industry will be earlier than the one for the optimal solution. If biological control is already available, the gains from government intervention in the competitive system may not be substantial enough to cover its administrative costs since the losses occur only when the existing technology is used. However, intervention is likely to be more profitable when biological control is not yet available and needs to be developed. In this case the susceptibility level will be exhausted sooner in the competitive solution, producing additional losses to society until biological control is ready to be utilized. If $S_w < 0$, government intervention is more desirable since, by proposition 4, the competitive solution through higher final levels of resistance reduces the future benefits of the alternative technology.

One then faces the problem of evaluating the costs of resistance† which would determine (1) the gains from government intervention and (2) the gains in developing new pest control technologies. The difference between the benefits of the optimal solution (ϕ^*) and the benefits of the private solution ($\hat{\phi}$) defines the cost of resistance due to the decentralization of the decision process. This sets an upper bound for the costs of government intervention to be economically valid. Considering a socially optimal program, the rate of investment in new pest control technologies simply should be equal to S_α, the marginal rate of return of this technology.

---

†Note that the total cost of resistance, however, which does not have any policy implications is $[\max_x \pi(x, w_o)]/r - \phi^*$.

The marginal cost of resistance (λ_t) indicates the value of a reduction in the resistance level given the optimal policy and available future alternative. Its time path, therefore, serves as a criterion for investment aimed at improving the current pest control practices by reducing their resistance level.

REFERENCES

Brown, A.W.A. (1967), Genetics of insecticide resistance in insect vectors, in *Genetics of Insect Vectors of Disease*, J.W. Wright and R. Pal, eds., Elsevier, Amsterdam, 505-552.

Georghiou, G.P. (1972), The evolution of resistance to pesticides, *Ann. Rev. Ecol. and Syst.*, 2, 133-163.

Gutierrez, A.P., J.B. Christensen, W.B. Loew, C.M. Merritt, C.S. Summers and W.R. Cothran (1976), Alfalfa and the Egyptian alfalfa weevil, *Can. Entomol.* (in press).

Hueth, D. and U. Regev (1974), Optimal agricultural pest management with increasing pest resistance, *Am. J. Agric. Econ.*, 56, 543-552.

Mardzhangan, G.M., Y.A.K. Ust, Z.S. Manukyan (1969), Population changes of the alfalfa weevil Phytonomus variablis Coleoptera, Curculionidae, in relationship to its resistance to BHC in the Araral Valley of Armenia, *Entomol. Obozrenie*, 48, 755-760.

Nordhaus, W.D. (1973), The allocation of energy resources, *Brookings Papers on Economic Activity*, 3, 529-576.

Regev, Uri, A.P. Gutierrez and Gershon Feder (1976), Pests as a common property resource: a case study of alfalfa weevil control, *Amer. J. Agric. Econ.*, 58, 186-197.

Scott, A.D. (1967), The theory of the mine under conditions of certainty, in *Extractive Resources and Taxation*, Gaffney, ed., University of Wisconsin Press, Madison, 25-62.

Taylor, C.R. and J.C. Headley (1975), Insecticide resistance and the evaluation of control strategies for insect populations, *Can. Entomol.*, 107, 237-242.

OPTIMAL MANAGEMENT OF AN ALFALFA ECOSYSTEM

Christine A. Shoemaker[1]

INTRODUCTION

A basic goal in pest management research has been the establishment of guidelines for suitable timing of pest control measures. A very important aspect of this research has been the establishment of economic thresholds which indicate when pest infestations are sufficiently severe to justify an insecticide treatment. However, as originally defined (Stern et al., 1959), the economic threshold is based only upon pest density. It is clear that in many situations the advisability of implementing a pest control measure also depends upon plant vigor and maturity, the density of natural enemy populations, on the weather and/or the age structure of the pest population.

It is, therefore, desirable to establish multi-dimensional economic thresholds, which provide pest management guidelines as a function of several variables, describing relevant environmental conditions and population densities. It is extremely difficult to establish such thresholds using purely empirical methods. The goal of this paper is to discuss the use of optimization methods to establish multi-dimensional economic thresholds and to describe the application of such methods to alfalfa weevil management.

Several papers have been published which utilize optimization methods to generate pest management guidelines. A summary of these is given in Table 1. The papers listed are classified by the number of populations and age-classes included, the types of control measures considered, and by the inclusion of genetic variability in the model. In the following sections we will discuss the model listed under number VII in Table 1.

This model is designed to estimate the best combination of biological, chemical and cultural methods for alfalfa weevil control. Alfalfa weevil populations can be reduced by the application of insecticide following harvest and by biological control provided by a larval parasitoid, Bathyplectus curculionis. Another important means of controlling alfalfa weevil is to harvest the hay early, during the peak of the population of small larvae. However, early harvesting can also decrease yield and decrease the size of subsequent parasite generations.

---

[1]Department of Environmental Engineering, Cornell University, Ithaca, New York 14853, U.S.A.

Table 1. Pest management models utilizing optimization methods.

Number	Populations Included	Number of Variables	Genetic Variation	Pest Management Methods	Optimization Method Used	Specific Pest Application	Reference
I	Pest	3	Yes	Insecticide	Discrete maximum principle	None	Hueth and Regev (1974)
II	Pest	3	Yes	Insecticide	Dynamic programming	None	Taylor and Headley (1975)
III	Pest	7	No	Chemo-therapy	Control vector iteration	Cancer	Bahrami and Kim (1975)
IV	Pest Parasite	1 1	No	Insecticide	Dynamic programming	*Anagasta kühniella*	Shoemaker (1973)
V	Pest Plant	3 1	No	Insecticide	Reduced gradient	Egyptian alfalfa weevil	Regev, Gutierrez and Feder (1976)
VI	Pest Plant	1 3	No	Insecticide Logging	Dynamic programming	Spruce budworm	Winkler (1975)
VII	Pest Parasite Plant	7 3 5	No	Insecticide Harvesting	Dynamic programming	Alfalfa weevil	See text

A DECISION MODEL FOR ALFALFA WEEVIL MANAGEMENT

The purpose of the management model is to determine the best times for harvesting and applying insecticide to alfalfa. The effects of insecticide depend upon the size and age structure of the population at the time of application. These in turn depend upon weather and the size of overwintering parasite and weevil populations. The effect of time of harvest on yield and subsequent weevil and parasite densities also depends upon these factors. Thus, we wish to determine the best times of harvesting and insecticide treatments as a function of weevil density, parasite density and weather.

Dynamic programming, the optimization method used in this model, cannot be used for problems with a large number of state variables. Unfortunately, an adequate model of an alfalfa ecosystem would require more than sixteen state variables: one for each age-class of the weevil and parasite populations, one for weather and several variables describing different parts of the alfalfa plant.

The difficulties associated with the large dimension of the alfalfa system can be circumvented by decomposing the pest management model into two coupled models. The first is the <u>population model</u>, which calculates the amount of feeding damage and the number of parasites and weevils reaching the overwintering stage as a function of Z_1, Z_2, Z_3, t_h and v where

Z_1 - number of adult weevils in the spring
Z_2 - number of adult parasites in the spring
Z_3 - the weather pattern
t_h - time of harvest
v - amount of insecticide applied after harvest.

Whenever necessary for clarity, the superscript n will be used with Z_1, Z_2, Z_3, t_h or v to denote "in the nth year".

The results from the population model are incorporated into three equations G_1, G_2, and D where

$$Z_1^{n+1} = G_1(Z_1^n, Z_2^n, Z_3^n, t_h^n, v^n) \quad . \tag{1}$$

$$Z_2^{n+1} = G_2(Z_1^n, Z_2^n, Z_3^n, t_h^n, v^n) \quad . \tag{2}$$

Thus G_1 and G_2 calculate the effects of cutting time, t_h^n and insecticide application v^n and the state variables Z_1^n, Z_2^n, and Z_3^n on the size of subsequent populations of weevils (Z_1^{n+1}) and parasites (Z_2^{n+1}).

The population model also calculates the function D which is the amount of alfalfa yield lost due to alfalfa weevil feeding. The value of D also depends upon Z_1, Z_2, Z_3, t_h, and v. The actual yield is

$$Y(Z_3, t_n) - D(Z_1, Z_2, Z_3, t_h, v) \tag{3}$$

where Y is the yield expected in the absence of weevil feeding with a first harvest at time t_h and a weather pattern Z_3.

The second model, which shall be referred to as a <u>decision model</u> is an optimization model which chooses the best time of harvest t_h^n and insecticide treatment v^n in year n as a function of Z_1^n, Z_2^n, Z_3^n. The best management policies are assumed to be those which maximize net income. The net income over N years is then

$$\sum_{n=1}^{N} \alpha^n \left[p^n \left(Y(Z_3^n, t_h) - D(Z_1^n, Z_2^n, Z_3^n, t_h^n, v^n)\right) - C^n(v^n) \right] \tag{4}$$

where p^n is the price per ton of alfalfa, C^n is the cost of insecticide treatment v^n, and α^n is the discount rate. Our goal is to maximize the expression in equation (4). The values of Z_1^n and Z_2^n are determined by the functions G_1 and G_2 in equations (1) and (2). The value of Z_3 is assumed to fluctuate stochastically. Its probability distribution is described by

$$\rho_i = P\,[Z_3 = \eta_i] \ . \tag{5}$$

i.e., ρ_i is the probability that the weather pattern is η_i, $i = 1,2,\ldots m$.

Formulating the problem of maximizing (4) subject to the constraints of equations (1), (2) and (5) as a dynamic programming problem results in the following equation:

$$F^n(Z_1{}^n, Z_2{}^n, Z_3{}^n) = \text{Max}_{t_h{}^n,\, v^n} [\alpha^n p^n(Y^n(Z_3{}^n, t_h{}^n) - D(Z_1{}^n, Z_2{}^n, Z_3{}^n, t_h{}^n, v^n)) - \alpha^n C^n(v^n) + \sum_{i=1}^{m} \rho_i F^{n+1}(Z_1{}^{n+1}, Z_2{}^{n+1}, \eta_i)] \quad (6)$$

for

$$Z_1{}^{n+1} = G_1{}^n(Z_1{}^n, Z_2{}^n, Z_3{}^n, t_h{}^n, v^n)$$
$$Z_2{}^{n+1} = G_2{}^n(Z_1{}^n, Z_2{}^n, Z_3{}^n, t_h{}^n, v^n) \quad (7)$$

By applying the dynamic programming algorithm iteratively to equations (6) and (7), we obtained the optimal choices of t_h and v which are presented in Figures 1-3. The calculations were possible because the optimization problem was reduced to one of three state variables Z_1, Z_2 and Z_3. Complicated relationships among weather, parasite-host synchrony, time of harvest, etc., are all incorporated in the functions G_1, G_2 and D. A description of the population model which calculates G_1, G_2 and D is given in the next section.

POPULATION MODEL

The length of time a weevil requires to complete its development depends upon temperature. Since the elapse of calendar time cannot be used to determine the physiological age of an individual, it is useful to define a new variable s, which is called physiological time. The variable s is defined as the cumulative number of units of heat which contribute to growth. Since R(t) measures the rate of development, the physiological time s corresponding to calendar time t is defined to be

$$s = g(t) = \int_0^t R(t)dt \quad (8)$$

where R(t) is a non-linear function of temperature estimated by Ruesink (1976). Equation (8) converts from units of calendar time t into units of physiological time s. It will sometimes be necessary to convert in the opposite direction, from units of physiological time s into units of calendar time t. Since $R(t) \geq 0$, Equation (8) defines a one-to-one correspondence between s and t. Thus, the inverse function, which we shall call h, is well-defined and

$$t = h(s) = g^{-1}(s) \quad . \quad (9)$$

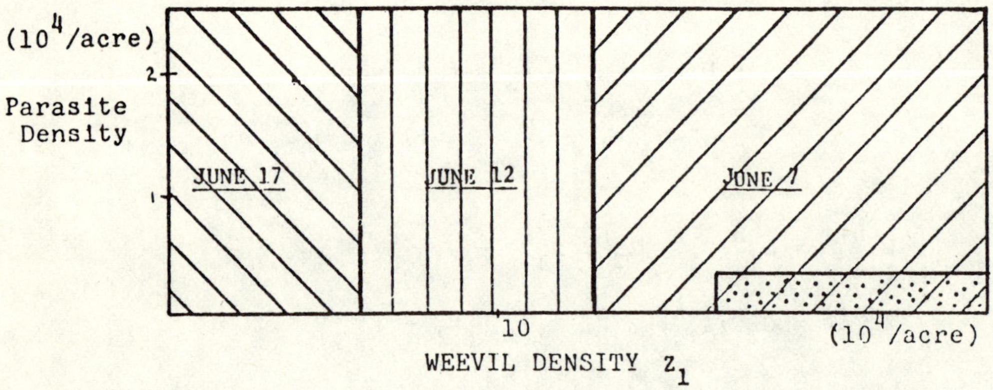

Figure 1. Multi-dimensional economic thresholds for cool weather and a one year planning horizon (Z_3=1, N=1).

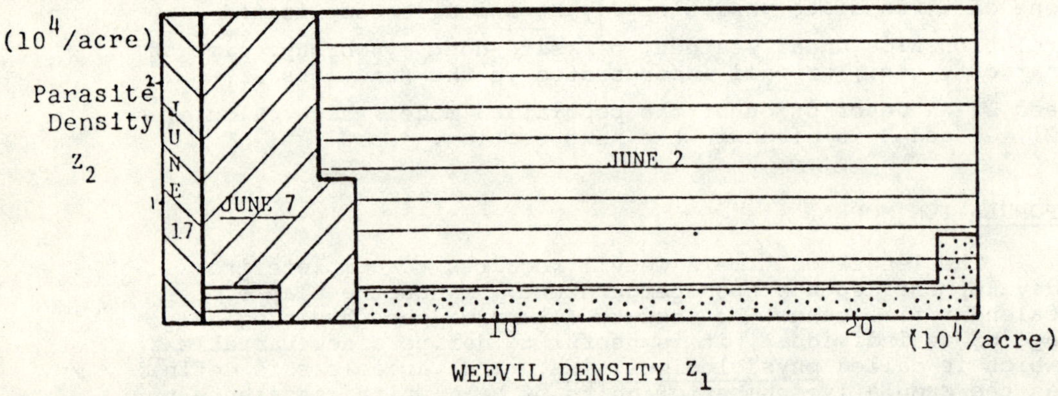

Figure 2. Multi-dimensional economic thresholds for warm weather and a three-year planning horizon (Z_3=3, N=3).

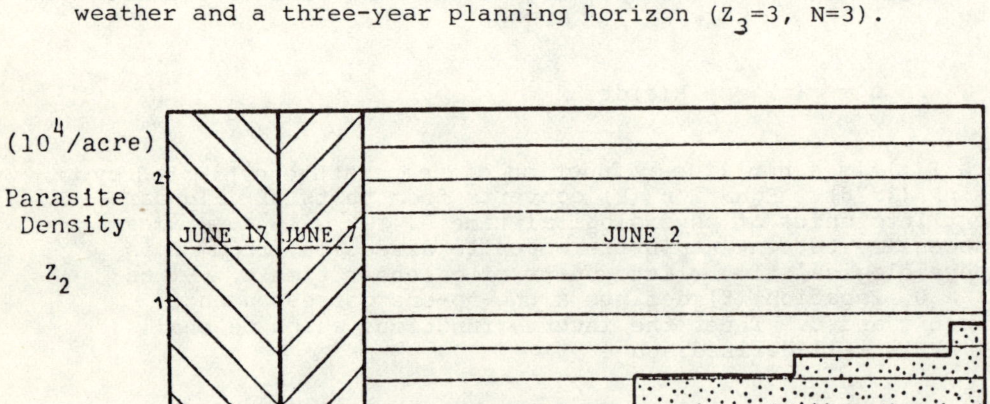

Figure 3. Multi-dimensional economic thresholds for warm weather and a one-year planning horizon (Z_3=3, N=1).

Since the rate of development R(t) depends upon the temperature T(t), the functions g(t) and h(s) depend upon the state variable Z_3^n which describes the temperature pattern in year n. The values for Z_3 were obtained by calculating the mean and standard deviation of the physiological time s(t) from six years of weather data for Ithaca, New York.

From the definition of physiological time, it follows that an individual oviposited at b has a physiological age s-b at physiological time s. Let N(s,b) be the number of weevils alive at physiological time s which were oviposited at physiological time b. Let A_i be the age at which a weevil enters the ith age-class. Thus, if $A_i \leq s-b \leq A_{i+1}$ then all of the individuals N(s,b) are in the ith age-class. Age-class zero represents eggs. Age-classes one through four represent the first through the fourth instars. Age-class five represents the pupal stage, and age-class six represents the adult stage.

Parasitism

The rate of parasitism at time t depends upon the number of female parasites P(t) which are searching for and attacking weevil larvae at time t. Adult parasites do not all emerge at the same time. Thus, the number of ovipositing parasites will vary with time.

Let $P_o(t)$ be the fraction of the total parasite population Z_2^n which are ovipositing females at time t. Then

$$P(t) = \begin{cases} Z_2^n P_o(t) & \underline{p} < t < \bar{p} \\ 0 & \text{otherwise} \end{cases} \tag{10}$$

where $(\underline{p},\bar{p})$ is the period of time during which the parasites are attacking alfalfa weevils.

Let H'(N(s,b), P) define the rate at which weevils of age s-b are parasitized. We will assume that H' is proportional to the number of weevils and that it is zero for weevils not in the susceptible age range (A_2, A_3). Thus,

$$H'(N(s,b), P) = \begin{cases} N(s,b) \cdot H(P) & A_2 \leq s-b \leq A_3 \\ 0 & \text{otherwise} \end{cases} \tag{11}$$

for some function H. The function H is defined in terms of calendar time, i.e., it is the number of prey attacked per unit of calendar time. To convert H(P) to units of physiological time we must multiply it by h(s).

A cohort of weevils N(s,b) is reduced by mortality from natural causes as well as by parasitization so

$$\frac{dN(s,b)}{ds} = -\mu(s-b)N(s,b) - H(P(h(s)))\,\dot{h}(s)\,N(s,b) \qquad (12)$$

where μ(a) is the rate of mortality from natural causes for weevils of age a. The parasitized population is decreased by death and increased by parasitization. Thus,

$$\frac{dN_p(s,b)}{ds} = -\mu(s-b)N_p(s,b) + H(P(h(s)))\,\dot{h}(s)\,N(s,b)\,. \qquad (13)$$

The initial conditions for equations (12) and (13) are

$$\begin{aligned} N(b,b) &= Z_1{}^n\,\theta(b) \\ N_p(b,b) &= 0 \end{aligned} \qquad (14)$$

where θ(b) is a normalized rate of weevil oviposition through time. The solutions to equations (12), (13) and (14) are

$$N(s,b) = Z_1{}^n\,\theta(b)\,e^{-(K(s-b) + L(s,b,P))} \qquad (15)$$

$$N_p(s,b) = Z_1{}^n\,\theta(b)\,e^{-(K(s-b))}(1-e^{-L(s,b,P)}) \quad \text{for } s-b<A_5 \qquad (16)$$

where

$$K(a) = \int_0^a \mu(r)dr \qquad (17)$$

and

$$L(s,b,P) = \int_{\underline{t}(s,b)}^{\bar{t}(s,b)} H(P(t))dt\ . \qquad (18)$$

The interval $(\underline{t},\bar{t})$ is the period of time before time s, when individuals oviposited at time b are susceptible to predation. Since weevils are assumed to be susceptible only while they are second instars, i.e., from age A_2 to age A_3,

$$\underline{t}(s,b) = h\ (\mathrm{Min}(s,b+A_2)) \tag{19}$$

$$\bar{t}(s,b) = h\ (\mathrm{Min}(s,b+A_3))\ . \tag{20}$$

MORTALITY FROM HARVESTING AND INSECTICIDE

As mentioned earlier, weevil mortality during harvest is age-specific. Mortality is much higher among younger forms than among more mature ones. Insecticide effectiveness is also age-specific but the older forms are most vulnerable. Let $\phi(v,a)$ be the fraction of weevils of age a which survive a harvest when v insecticide is applied. Insecticide applications are assumed to occur just after harvest, if at all. Since t_h is the calendar date of harvest, $s_h = g(t_h)$ is the physiological time of harvest. Thus, a weevil oviposited at time b is of age s_h-b at the time of harvest.

Thus, for $s>s_h$ the number of individuals alive at time s who were born at time b is the product of $\phi(v,s_h-b)$ and the number which would be alive if no harvest mortality took place. The latter value is given in equations (15) and (16). Thus, after harvest

$$N(s,b) = \phi(v,s_h-b)Z_1{}^n\theta(b)e^{-[K(s-b)+L'(s,b,P)]} \tag{21}$$

$$N_P(s,b) = \phi(v,s_h-b)Z_1{}^n\theta(b)e^{-K(s-b)}[1-e^{L'(s,b,P)}] \tag{22}$$

for $s>s_h$.

The function L has been replaced with the function L' which incorporates the effect of harvesting and insecticide treatments on the number of parasites.

Yield

In order to solve the decision model described in equation (6) it is necessary to estimate both the yield in the absence of weevil feeding $(Y(Z_3,t_h))$, and the amount of yield loss caused by alfalfa weevil ($D(Z_1,Z_2,Z_3,t_h,v)$). The yield $Y(Z_3,t_n)$ is calculated from a simulation model of alfalfa growth (ALSIM) developed by Fick (1975).

Alfalfa weevil feeding damages the crop in two ways. The first is to remove leaf tissue and thereby reduce yield. Let $f(a)$ and $f_p(a)$ be the rate of feeding by unparasitized and parasitized alfalfa weevils, respectively, of age a. Then total leaf feeding equals

$$\int_0^{\bar{b}} \int_b^{b+A} [f(s-b)N(s,b) + f_p(s-b)N_p(s,b)]ds\, db \, . \qquad (23)$$

Equation (23) can be solved by substituting equations (15) and (16) (for $s<s_h$) and equations (21) and (22) (for $s>s_h$) for the values of $N(s,b)$ and $N_p(s,b)$.

Besides removing leaf material, alfalfa weevil feeding can also delay regrowth of the alfalfa after it has been cut. Immediately after harvest the surviving weevils have no leaves to feed upon so they feed instead upon the buds. Although the total mass of material removed by this feeding may be small, its detrimental effects are significant. Fick (1976) has shown that weevil feeding on buds can delay the start of regrowth by as much as ten days. He has also shown that for regrowth delays of less than seven days, the length of the delay is a linear function of larval density.

Therefore, for the alfalfa weevil management model it is assumed that

$$\text{delay} = k[\int_o^{\bar{b}} [f(s_h-b)N(s_h,b)+f_p(s_h-b)N_p(s_h,b)\, db]] \qquad (24)$$

where s_h is the physiological time of the harvest and $N(s_h,b)$, $N_p(s_h,b)$, $f(a)$ and $f_p(a)$ are defined as above. The value of the constant k was obtained from Fick (1976). The effect of the delay on total yield was estimated from simulations of Fick's ALSIM model for each cutting time and weather pattern. The function D is then the sum of the yield loss due to the delay in regrowth (equation 24) plus the loss due to leaf feeding given in equation (23).

Overwintering Populations

Weevil oviposition is the result of mating primarily in the early spring by individuals who overwintered as adults. The number of such adults in the spring of the n+1th year, z_1^{n+1} is the number of weevils who completed their life cycle in the summer of the nth year, multiplied by the rate λ_w at which they survive the winter. Thus,

$$z_1^{n+1} = \lambda_w \int_o^{\bar{b}} N(b+A_7,b)\,db \tag{25}$$

where $\bar{b}$ is the end of oviposition. By substituting Equation (21) for $N(b+A_7,b)$, we obtain

$$\begin{aligned} z_1^{n+1} &= \lambda_w z_1^n e^{-K(A_7)} \int_o^{b} \theta(b)\phi(v,s_h-b)\,e^{-L(b+A_3,b,P)}\,db \\ &= G_1(z_1^n,z_2^n,z_3^n,t_h^n,v^n) \quad . \end{aligned} \tag{26}$$

Total oviposition by parasites is assumed to be proportional to the number of overwintering parasites. Thus, the total number of adult parasites in year n+1 is

$$z_2^{n+1} = \lambda_p \int_o^{\bar{b}} N_p^n(b+A_5,b)\,db \tag{27}$$

where λ_p is the overwintering survivorship of the parasites. Substituting equation (22) into equation (27), we obtain

$$\begin{aligned} z_2^{n+1} &= \lambda_p z_1^n e^{-K(A_5)} \int_o^{\bar{b}} \theta(b)\,[1-e^{-L'(b+A_3,b,P)}] \\ &\phi(v,s_h-b)\,db = G_2(z_1^n,\ z_2^n,\ z_3^n,\ t_h^n,\ v^n) \quad . \end{aligned} \tag{28}$$

The values of the functions G_1 and G_2 from equations (26) and (28) are substituted into equations (6) and (7) of the decision model to obtain the optimal times for harvesting and insecticide treatments.

RESULTS

Equation (28) completes the alfalfa weevil management model. The decision model is described in equations (6) and (7). The effects of management on population sizes are described in equations (26) and (28) by functions G_1 and G_2. The parameter values in the population model are based on field data collected in central New York by Robert Helgeson, Department of Entomology, and Gary Fick, Department of Agronomy, Cornell University. The only remaining unknowns in equations (6) and (7) are the values of t_h^n, v^n and F^n. These values are calculated by the application of a dynamic programming algorithm.

The results of computations utilizing this algorithm are shown in Figures 1-3. Since the optimal policies calculated depend upon three variables—alfalfa weevil density, parasite density and weather—the results presented in Figures 1-3 describe multi-dimensional, economic thresholds for alfalfa weevil management.

The optimal management policy has two components, the time of harvesting and the amount of insecticide applied. The patterns appearing in Figure 1 correspond to date of the first harvest which range from June 1 to June 17. The amount of insecticide applied is either zero or the standard dosage. Dots mark those combinations of weevil and parasite densities for which insecticide treatments are calculated to be part of an optimal policy.

The results presented in Figure 1 are for a one-year planning horizon, i.e., for N equal one. Thus, the calculations do not include the effect of the current year's pest control policies on pest densities in subsequent years. This is an appropriate assumption if the area of alfalfa being managed is infested primarily by an uncontrollable migration of adult alfalfa weevils and B. curculionis from outside areas.

In the absence of weevils, the alfalfa simulation model estimates that the maximum yield is obtained by harvesting the first growth at the last possible date, June 17. However, as weevil density increases, the decision model calculates that the optimal harvesting time becomes earlier. Early harvesting is calculated to be preferable to insecticide treatments in most cases. Only when weevil densities are very high, and parasite densities are very low, are insecticide treatments recommended.

The weather pattern assumed for Figure 2 is warmer than for Figure 1. This difference affects both the rate of insect development and plant growth. For moderate to high weevil densities, the optimal time of harvest is earlier for average weather than for cool weather. This change in optimal policies occurs because the warmer temperatures in Figure 2 cause the weevil, as well as the crop, to develop more quickly. Thus, there is more of a need for early weevil control. In addition, an early harvest causes less of a reduction in total yield because the crop matures more quickly with warmer temperatures.

The pest management policies calculated to be optimal for a three-year planning period and warm weather are presented in Figure 3. Because of the longer-term planning horizon, the effect of management decisions on population levels in subsequent years influences the choice of harvest date and insecticide treatment. As a result, the optimal results presented in Figure 3 recommend a more intense control of the alfalfa weevil: harvest dates are generally earlier and insecticide treatments more frequent than in Figure 3.

To obtain the optimal policies for all values of Z_1^n, Z_2^n, and Z_3^n for one, two and three-year planning horizons from the model described, cost less than $12 on an IBM 370-168 computer. The economy of these calculations is based upon an algorithm developed in Shoemaker (1976). The low cost of obtaining the results in Figures 1-3 illustrates the advantage of using an optimization method over exhaustive simulation to choose the best management program from a large number of options.

To obtain the same results by simulation would involve the calculation of a three-year model of insect and plant growth for each combination of values of Z_1^1, Z_2^1, Z_3^1, Z_3^2, Z_3^3, t_h^1, t_h^2, t_h^3, v^1, v^2, v^3. In the above model, we consider ten values of Z_1^1 and Z_2^2, three values of Z_3^1, Z_3^2, Z_3^3, four values of t_h^1, t_h^2, t_h^3, and two values of v^1, v^2, v^3. Thus the total number of necessary simulations would be

$$10 \times 10 \times 3 \times 3 \times 3 \times 4 \times 4 \times 4 \times 2 \times 2 \times 2 = 1.38 \text{ million.}$$

The cost of such a large number of simulations makes simulation an impractical method for calculating the best policies for alfalfa weevil management over a wide range of population densities and weather conditions.

The algorithm developed for decomposing the alfalfa weevil model can be directly applied to management models for other pests provided:

a) the pest population has discrete generations,
b) pest control measures are applied only once per generation,
c) instantaneous feeding damage is proportional to the number of pests,
d) the instantaneous rate of parasitization is proportional to the number of hosts.

The algorithm can be modified to include situations which do not satisfy some of the above conditions. However, the calculations may become considerably more expensive. On the other hand, if the population model does not include a parasite or predator, the calculations can be expected to be much less expensive than for the alfalfa weevil management model.

ACKNOWLEDGEMENTS

Research on the application of dynamic programming to insect pest management was supported in part by a grant from the National Science Foundation Engineering Research Initiation Grant GK42147. This grant also supported the work of Albert Valocchi and Leland Baskin who were responsible for computer programming.

Most of the parameters in the alfalfa weevil model were based on field studies by Robert Helgeson and Gary Fick of Cornell University, on a literature review by William Ruesink of the Illinois Natural History Survey, and on statistical analyses by Gary Smith and the author. All of the above work was supported by a grant to the University of California from the National Science Foundation and the Environmental Protection Agency (GB 34718).

The findings, opinions and recommendations expressed herein are those of the author and are not necessarily those of the University of California, the National Science Foundation or the Environmental Protection Agency.

REFERENCES

Bahrami, K. and M. Kim (1975), Optimal control of multiplicative control system arising from cancer therapy, *IEEE Trans. Auto Control*, 537-542.

Fick, G.W. (1975), ALSIM 1 (Level 1) User's Manual, Cornell University, Department of Agronomy, Mimeo, 75-20.

Fick, G.W. (1976), Alfalfa weevil effects on regrowth of alfalfa, *Agronomy Journal*, 68, 809-812.

Hueth, D. and U. Regev (1974), Optimal agricultural pest management with increasing pest resistance, *Amer. J. Agr. Econ.*, 56, 543-551.

Regev, U., A.P. Gutierrez, and G. Feder (1976), Pests as a common property resource: a case study of alfalfa weevil control, *Amer. J. Agr. Econ.*, 58, 188-195.

Ruesink, W.G. (1976), Modeling of pest populations in the alfalfa ecosystem with special reference to the alfalfa weevil, in *Modeling for Pest Management*, R.L. Tummala,

D.L. Haynes, and B.A. Croft, eds., Michigan State University Press, East Lansing, Michigan, U.S.A.

Shoemaker, C.A. (1973), Optimization of agricultural pest management III: Results and extensions of a model, *Mathematical Biosciences*, 18, 1-22.

Shoemaker, C.A. (1976), *Optimal Integrated Pest Management of Age Distributed Populations*, Department of Environmental Engineering Technical Report 76-2, Cornell University.

Stern, V.M., R.F. Smith, R. van den Bosch, and K.S. Hagen (1959), The integrated control concept, *Hilgardia*, 29, 81-101.

Taylor, C.R. and J.C. Headley (1975), Insecticide resistance and the evaluation of control strategies for an insect population, *Can. Entomol.*, 107, 237-242.

Winkler, C. (1975), *An Optimization Technique for the Budworm Forest-Pest Model*, International Institute for Applied Systems Analysis, Laxenburg, Austria, RM-75-11.

INSTITUTE OF FOREST AND WOOD,
SIBERIAN BRANCH OF THE ACADEMY OF SCIENCES OF THE U.S.S.R.

INERTIAL AND NONINERTIAL FACTORS REGULATING FOREST INSECT POPULATION DENSITY

A. S. Isaev, R. G. Khlebopros

In developing a theory of population dynamics, of essential interest are general principles of cybernetics, describing mechanisms of system self-regulation irrespective of the physical nature of the systems. These principles are widely used in ecological research (Solomon, 1957, 1964; Wilbert, 1961; Viktorov, 1963, et al.), and make it possible to consider forest insect population dynamics as a self-regulating process which is governed by a complex of natural factors operating according to the principle of negative feedback.

The concept of insect number self-regulation is generally accepted now. This concept has led to the synthetic theory of population dynamics (Schwerdtfeger, 1968), a theory which has won wide recognition. However, until recently, this theory lacked a good mathematical foundation which would permit a description of the process of self-regulation of dynamic ecological systems. Systems analysis of forest insect population dynamics has led to the development of a method which allows one to discriminate between modifying and regulating effects of environmental factors (Isaev, Khlebopros, 1973, 1974).

The new method of analysis of forest insect population dynamics proposed here is based on the principle of stability in dynamic ecological systems. Forest biogeocenosis as a self-contained ecological system in any of its parameters will have a domain of stability in which the system is not subject to any significant changes. Different rates of motion of individual biogeocenosis components permit one to single out certain bits of the system, i.e., to distinguish groups of populations which respond to the effects of environmental factors at a similar rate. This approach is rather promising as it permits an analysis of variations in population density with relative stability of the entire system. The analysis of the forest insect population in a phase plane permits us to reveal the most typical points of the system, the shift from one point to another being associated with changes in forest insect population dynamics. As successions of forest biogeocenoses are incommensurable in time with the period of insect generations, the typical points of the dynamics of a forest insect population can be regarded as fixed in a phase plane relative to variations in insect numbers.

Using a functional relation between population density and reproductive rate, we found the following typical points of forest insect population dynamics: i) the stable density (x_1) of a scarce population at which a species generally occurs in a stable biogeocenosis; ii) the threshold density (x_R) which once exceeded allows a species to develop an outbreak; iii) the optimum density corresponding to maximum reproductive rate (x_c); iv) the upper density (x_T) which is characteristic of the peak phase of an outbreak; v) the critical density (x_o) which is responsible for qualitative changes in a population state; vi) the metastable high density (x_2) associated with long-term destructive effects of insects on forest biogeocenoses.

Though the presence of negative feedback is a necessary condition for the existence of all self-contained systems, the efficiency of regulation is determined by inertial possibilities of regulating factors, i.e. by the period of time lag of these factors to respond to variations in population density. It seems reasonable, therefore, to divide factors regulating a population density into inertial with a lag, and noninertial whose lag is insignificant or absent. The first category includes biocenotic factors (natural enemies, diseases, food plant response), while the second one includes most of the intraspecific regulating factors (competition, interference, cannibalism, migrations etc.). This division is of great importance as it allows a different view of the nature of regulation of forest pest numbers under the conditions of a given biogeocenosis.

In our recent work we have dealt with a phase trajectory of the population density of certain forest insect species and analyzed how regulating factors operate in the domains of stability and outbreak (Isaev, Khlebopros and Kondakov, 1974). Our further studies have demonstrated that the typical points of the system together with individual estimation of regulating factors, according to the degree of their inertiality, might be used to construct a phase portrait of a population which would reflect peculiarities of the ecology of a species and its interactions in the forest biogeocenosis. The construction of an ecological phase portrait will make it possible to analyze population dynamics qualitatively and quantitatively, to elucidate the nature and significance of individual regulating factors, and to define conditions under which a species will be able to develop into an outbreak.

INERTIAL AND NONINERTIAL REGULATING FACTORS

Modern concepts concerning the dynamics of a forest insect population indicate that factors responsible for changes in population density can be divided into modifying and regulating ones (Viktorov, 1963). The effect of modifying factors (food, weather conditions) does not depend on population density and is of a stochastic nature. Regulating factors (natural enemies, diseases, intraspecific factors, etc.) counteract these deviations on the feedback principle and are dependent on population density. These factors act in an ecosystem with a certain lag relative to the number of a given species.

A number of papers have reported that the effect of a lag must be taken into account while analyzing population density. (Volterra, 1931; Kostitsin, 1937, Smith, 1970; Hitchinson, 1948; Viktorov, 1967). For our purposes it was interesting to consider the effect of a lag which is an intrinsic feature of self-contained ecological systems.

In an analysis of the self-regulation process in the phase plane y,χ where y is the reproductive rate and χ is population density, the value of a lag (τ) may be defined by the relaxation time of the regulating factors. The presence of a lag is due to the fact that the regulating effects of biotic factors are determined by the density of previous generations rather than the density of the present generation. If the population density within several generations is assumed to be unchanged then a lag is absent ($\tau=0$), hence $y=y_o$. Since population density in the forest biogeocenosis fluctuates widely the reproductive rate y will not be equal to y_o. Thus the difference $y-y_o$ defines the lag of biotic factors in relation to numbers of a given species. When the population becomes scarce, it is markedly affected by biotic factors (a negative lag); when population density is rapidly increasing, the effect of biotic factors in a particular period is reduced (a positive lag).

The phenomenon of a lag is a typical feature of each regulating factor, the value of the variable τ, however, can be essentially different for each factor. If the relaxation time is much shorter than the generation period of a given species ($\tau \ll T$) this regulating factor can be considered to act without lag. When the period of time lag and generation period are, or nearly, equal to each other ($\tau \simeq T$), the regulating factors are thought to have a certain period of time lag. The qualitative nature of insect population density and the nature of the stability of a forest biogeocenosis depend on which of these regulating factors govern the system.

CHARACTERISTICS OF MAJOR PARAMETERS OF INERTIAL AND NONINERTIAL REGULATING FACTORS

In order to perform an analysis of an insect population in the χ,y phase plane, let us consider the reproductive rate y as a function associated with population density χ, food quantity-quality (A), weather conditions (P) and a lag of regulating factors in relation to the numbers of a given species:

$$y=y(A,P,\chi,\tau) \quad . \qquad (1)$$

To elucidate the role of inertial and noninertial regulating factors, modifying factors must be regarded as being relatively constant. This assumption is of great importance, otherwise the phenomenological analysis of a phase portrait of a population is not feasible.

As different regulating factors produce a different effect on insect population dynamics, the problem is to measure the degree of their inertiality. It is most likely that biotic regulating factors (associated with the effect on a phytophage of its natural enemies—entomophages) are most inertial as generation periods of both populations are similar. Causative microorganisms are low-inertial factors when they work as "living insecticides". In natural situations, an epizootic is considerably inertial as infectivity of a population is markedly dependent on the ratio of sick to healthy individuals in previous generations. It should be noted that epizootics occur mainly at high population densities. The spread of infection is associated not only with a high host population density but also with the weakness of an organism due to a shortage of food. Under such circumstances, latent carriers of viruses begin to display great activity (Gershenson, 1959; Janisch, 1962, et al.), and microorganisms available in the environment may infest insects (Steinchaus, 1958).

Intraspecific regulating factors are also distinguished by the degree of their inertiality. The first group includes factors whose response is immediate. Such factors as competition, cannibalism, interference, changes in fecundity and migration reduce population density within a period of one generation. These factors have no period of time lag as the effect of regulation is attained within one generation and is almost independent of the density of the previous generation. All this tends to indicate that a lag is virtually absent. The second group includes factors which affect not only the population density of the present generation but also predetermine possible changes in the next one (diapause, changes in age composition, a situation when part of a population shows an accelerated development, genetic and phenotypic polymorphism, the sex ratio). The second group also includes factors of population growth (Naumov, 1963). These factors alter the behavioral responses of insects and their physiological states, resulting in a decline in numbers. A typical example of this kind of self-regulation which is not associated with competition is the phase variability reported for some Lepidoptera (Iwac, 1962; Sharov, 1959). Numerous studies have demonstrated that the phase variability of some insects depends on population density, and is an effective factor regulating insect numbers.

Consider the effect of a lag in a situation when forest insect numbers are regulated by the entomophages. Let the density of an entomophage be z, then

$$y=y(\chi,z) \qquad (2)$$

Experiments demonstrate that the nature of the interaction between insects and their natural enemies does not depend on entomophage density alone, as one and the same number of entomophages exerts a different regulatory effect on a population at a different density of the host. Thus it appears more convenient to use in the analysis the value of food supply Θ, i.e. the number of host individuals per one entomophage

$$\Theta = \frac{\chi}{z}$$

$$\text{then } y=y(\chi,\Theta) \qquad (3)$$

The analysis of a phytophage population in the y, Θ phase plane is of great interest since the phytophage reproductive rate (y) can be considered as being dependent on entomophage density. Unlike the χ,y phase plane, this enables one to examine individually the regulatory effects of species or groups of species of natural enemies on host-prey numbers.

Figure 1 shows y as a function of Θ when χ is constant. This kind of a relationship $y=f(\Theta)$ was first observed by Yanovsky and Kiselev (1975) in bark beetle (IPS. Subelongatus Motsh). These workers showed the presence of two threshold values of food resource, Θ' and Θ''. At these points the nature of insect regulation is different in kind. They concluded that when food supply is poor (up to the first threshold Θ') whatever phytophage density, host numbers are rigidly controlled by entomophages. Between the first and the second thresholds $\Theta' \le \Theta \le \Theta''$ a regulatory effect of entomophages is rapidly decreasing, and at $\Theta > \Theta''$ entomophages fail to control a phytophage population. In this situation phytophage numbers will be controlled by intraspecific mechanisms.

Using the above considerations, consider the interplay between regulating factors which are distinguished by their inertiality. The effect of a lag is defined by two main parameters: by the value of τ and the relative contribution of this factor (β) to overall regulation. The greater the value of τ, and its relative contribution β, the wider is the range of possible values for the reproductive rate $y_\beta - y_\alpha$ (Figure 1). If τ or β tend to a zero, then $y_\beta - y_\alpha 0$, and the curve $y=f(\Theta)$ turns into a straight line parallel to the axis of the abscissa. If the regulatory effect of entomophages is absent ($\beta \to 0$) the curve turns into a straight line $y=y_\beta$; when an inertial regulating factor is changed into a noninertial one ($\tau \to 0$), the curve takes the form of a straight line parallel

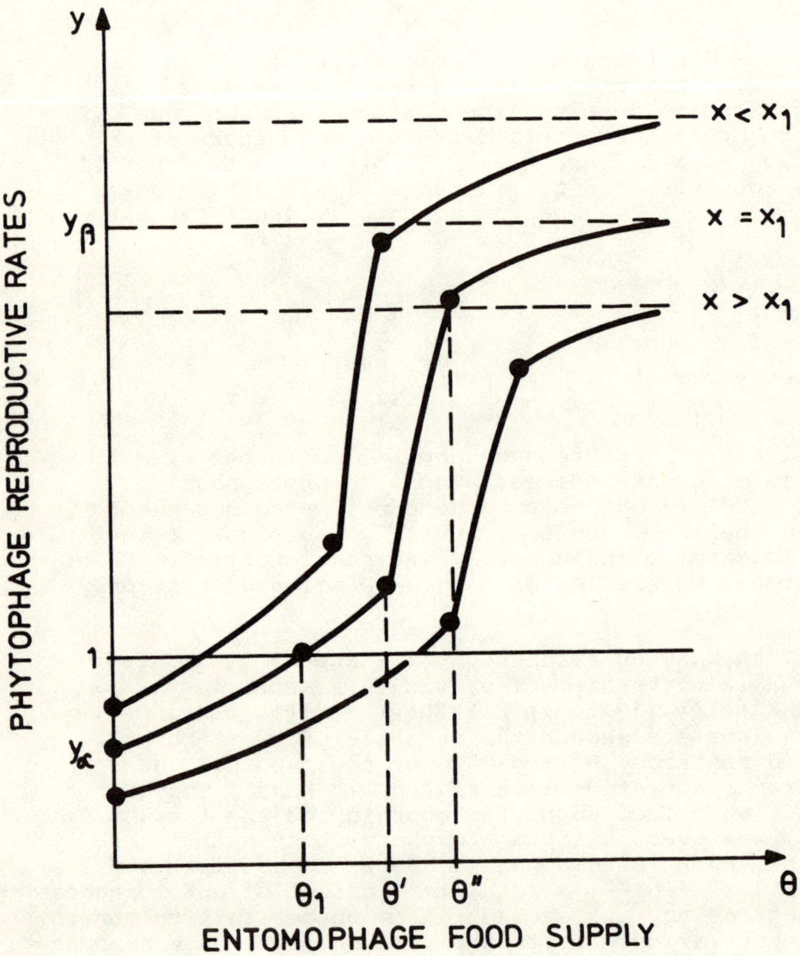

Figure 1. Phytophage reproductive rates as a function of food supply for their natural enemies

θ_1 - food supply at a stable population density (x_1)
θ' - the first threshold of food supply (at x_1)
θ'' - the second threshold of food supply (at x_1)

In the portion $0 \leq \theta \leq \theta'$ entomophages act as low-inertial factors; in the $\overline{\theta'} \leq \overline{\theta} \leq \theta''$ portion as inertial; in the $\theta > \theta''$ portion intraspecific factors are more important; $y_\beta - y_\alpha$ is the range of change in reproductive rates at x_1.

to the axis of the abscissa, and lies, depending on the relative contribution of this regulating factor, between y_α and y_β. If β is equal to unity, $y=y_\alpha$ (Figure 1). Thus, when fluctuations of modifying factors are small, the straight lines $y=y_\alpha$ and $y=y_\Theta$ in the y,Θ phase plane will represent boundary values which restrict the range of allowed values for the reproductive rate in this system.

Experiments state that the effect of entomophages in different portions of the $y-y(\Theta)$ curve is characterized by the derivative $\frac{\partial y}{\partial \Theta}$. The less the value of the derivative, the less inertial is the regulating factor and, hence, the more rigid is regulation (i.e. this factor does not cease to control population density). In the portion $0 \leq \Theta \leq \Theta'$ (up to the first threshold of food supply) entomophage numbers are so high ($Z \to \infty$) that at this X value the reproductive rate is not changed markedly (Figure 1). This is due to the fact that when food supply is poor, insect numbers are controlled virtually by a limited and relatively constant number of entomophages which is comparable with phytophage numbers as the latter are discretely consumed. Numbers of prey attacked increase in direct proportion to the density of phytophages, i.e. in this instance a functional response of entomophages to prey density is most pronounced (Solomon, 1949). Therefore, up to the first threshold of food supply, entomophages act as regulating factors whose period of time lag is near zero, the degree of their inertiality $(\frac{\partial y}{\partial \Theta})$ being determined by the biological and ecological nature of the interaction between the host and its natural enemies in a particular biogeocenosis.

For example, the first threshold of food supply for parasites and predators feeding on bark beetles is about 40 larvae per individual (Figure 2). One predator is known to consume 1 to 10 bark beetle larvae during a day (Zinovyev, 1959; Nikityuk, 1951). Thus even if predator population density is very high, the number of individuals participating in consumption (and, consequently, the number of infective larvae) is virtually the same. It is clear that in this instance the period of time lag is near zero, and entomophages act as low-inertial regulating factors ($\tau < T$).

In the portion of the curve between the first and the second thresholds ($\Theta' \leq \Theta \leq \Theta''$) the regulating factor associated with entomophages is most inertial, i.e. the partial derivative $\frac{\partial y}{\partial \Theta}$ in this region achieves its maximum value. It should be noted, however, that even with a high reproductive rate of phytophages, regulation is still present but it tends to indicate that the nature of regulation has changed from non-inertial to inertial. That is why between the first and the second threshold of food supply a host is able to escape its natural enemies and develop an outbreak.

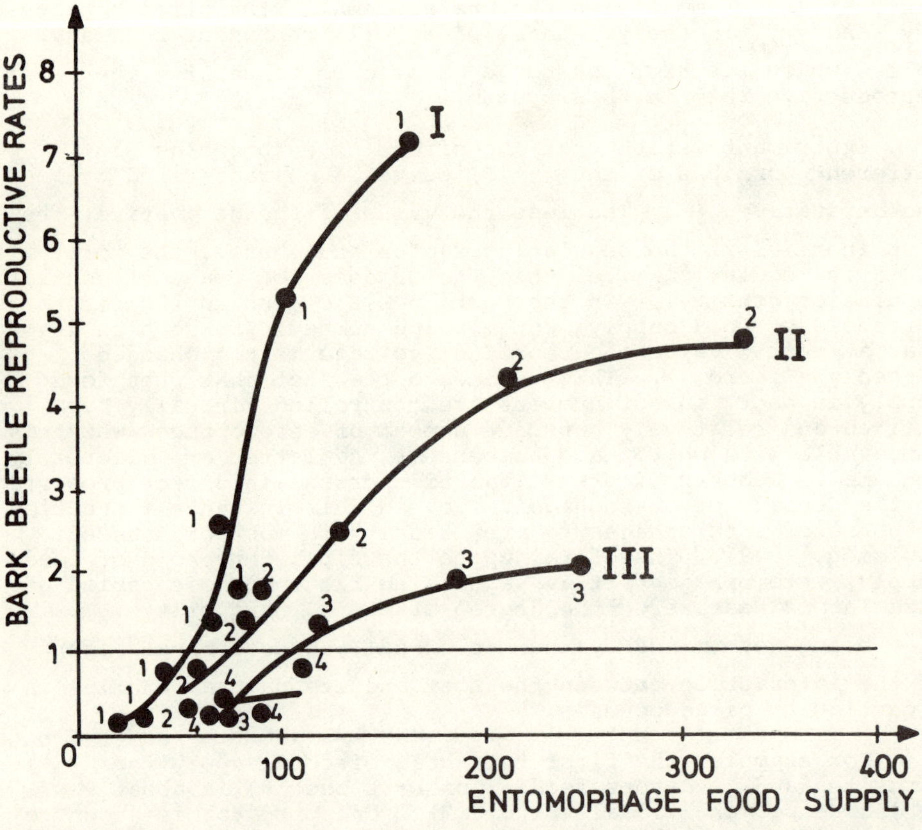

Figure 2. Bark beetle reproductive rates as a function of food supply for entomophages (according to Yanovsky and Kiselev, 1975).

Bark beetle densities (families per one dm^2): I - 0,48-0,77; 2 - 1,43-1,59; 3 - 2,33-2,52; 4 - 3,39-4,75.

An individual of any entomophagous species can consume only a limited number of host-prey. If the food available for entomophages is very close to their physiological limit of consumption, then, even with a slight change in entomophage numbers, phytophage reproductive rates may change markedly, allowing the latter to escape their natural enemies. In this portion, regulating pressure of entomophages is weaker due to a different qualitative manifestation of their functional and numerical responses. As in the portion $\theta' < \theta < \theta''$ the amount of available food exceeds consumption rates, the functional response of entomophages is reduced while the role of numerical response is raised. The numerical response of entomophages is characterized by a maximum increase in ration, a rise in fecundity and survival. A reduced functional response results in an increase in the derivative and, consequently, the regulating factor appears more inertial, while a rise in the numerical response tends to decrease the derivative $\frac{\partial y}{\partial \theta}$, and, hence, the regulating factor becomes less inertial. If changes in both functional and numerical responses fail to compensate each other, then even slight changes in a population density of natural enemies cause reproductive rates of the host-prey to increase rapidly. Under such circumstances, the negative feedback may change for a positive one, resulting in an outbreak. The ability of a phytophage to escape its entomophage under these conditions was first demonstrated by Tachacashi (Tachacashi, 1964). This process was summarized in our previous works (Isaev, Khlebopros, 1973, 1974). Let us now consider in more detail how highly inertial regulating factors generate positive feedback.

The stability of an ecological system is maintained by the presence of factors having negative feedback. The value of the feedback is defined by the full derivative

$$\frac{dy}{d\chi} = \frac{\partial y}{\partial \chi} + \frac{\partial y}{\partial \theta} \cdot \frac{\partial \theta}{\partial \chi} \qquad (4)$$

where $\frac{\partial \theta}{\partial \chi} = \frac{1}{z}$. Since we deal with a stable ecosystem, the value of the partial derivative $\frac{\partial y}{\partial \chi}$ must be negative, as seen in Figure 1. Experiments demonstrate that the curves $y=y(\theta)$ at χ_k = const. do not cross each other, and as χ is increasing, y is decreasing (Figure 2). The second term of Equation 4 is always positive, as the curve $y=f(\theta)$ is rising monotonically. Therefore, the derivative $\frac{\partial y}{\partial \theta}$ determines the sign of a feedback. A progressive rise in the derivative may lead to the inequality

$$\frac{\partial y}{\partial \theta} \geq \left|\frac{\partial y}{\theta \chi}\right| \cdot z \qquad (5)$$

and in the portion $\Theta' \leq \Theta \leq \Theta''$ of the $y=f(\Theta)$ curve there emerges positive feedback, i.e. when population density X increases, reproductive rates tend to increase as well. Under this situation insects are able to escape their natural enemies and develop an outbreak. The point Θ' can be regarded as the beginning of an outbreak, in which the negative feedback changes for the positive one, and there is a rapid increase in reproductive rates. At point Θ'' a rise in reproductive rates does not take place any longer, and negative feedback re-emerges.

The above considerations suggest that the effect of inertial factors makes the negative feedback weaker, and thus regulation becomes less rigid. Inertial factors of species capable of developing an outbreak play a relatively greater role in the regulating system than noninertial ones. It is known, for example, that scarce lepidoptera populations are controlled mainly by parasites. It was clearly demonstrated with the most dangerous pest of the taiga, Dendrolimus sibirica (Kolomeets, 1962; Rozhkov, 1965). The highly inertial regulating effect of entomophages on these species may be related to the presence of a complex set of entomophagous species, characterized by interspecific competition and great activity of secondary parasites. A convincing proof of this was provided by Zwolfer (1963) who studied a set of parasites for some Lepidoptera species at different population densities. Unlike species capable of developing an outbreak, Lepidoptera with a stable density have a limited set of specialized parasite species, the regulatory effect of which is, in fact, non-inertial. It is obvious that most forest insect species are controlled by low-inertial factors, i.e. their natural enemies "keep" a population density stable, thus preventing these species to develop an outbreak.

The regulatory effect of entomophages is virtually absent beyond the second threshold of food supply in the portion Θc. Food availability in this instance greatly exceeds consumption rates. For example, in the portion Θc the amount of food available for an entomophage feeding on bark beetle is different, depending on bark beetle population density, and ranges from 150 to 350 individuals per one entomophage (Figure 2). Hence, beyond the second threshold, the functional response of entomophages is absent, and the numerical response, having achieved its limit, does not affect host reproductive rates any longer. In this portion of the curve intraspecific factors, which act as noninertial ones, become progressively more important. In this portion, as the derivative $\frac{\partial y}{\partial \Theta}$ tends to zero, the $y=y(\Theta)$ curve is asymptotically approaching the value of $y=y_{\beta}$ (Figure 1).

It is significant that noninertial factors can function irrespective of the way in which the entomophage is supplied by food, although the effects of these factors become more important with an increase in the pest population. This results from the fact that when pest densities are high, other, new noninertial regulating factors are introduced in series into the overall regulating system, and thus the relative contribution of these factors increases. As seen in Figure 2, a rise in bark beetle population density leads to a decline in the threshold value and to a decreased steepness of the $y=y(\Theta)$ curve, indicating that the relative contribution of noninertial intraspecific factors becomes more important.

The proposed method analyzes in the y, Θ plane the projection of the relationships $y=y(\chi,\Theta)$ in three dimensional space, an experiment to be more informative, and by using it one can obtain the equation for the phytophage population dynamics.

It was stated previously (Isaev, Khlebopros, 1973, 1974) that

$$y = APe^{-\alpha\chi} \tag{6}$$

where α is the integral coefficient of the negative feedback. The equation

$$y = APe^{-\alpha_o \chi - \frac{\gamma Z}{\chi}} \tag{7}$$

approximates the $y=y(\Theta)$ curves obtained with xylophagous insects (Yanovsky, Kiselev, 1975), where α_o is the feedback coefficient of noninertial factors, and γ is a parameter of inertial factors. Hence, the relative growth rate of a population of xylophagous insects can be described by the equation

$$\frac{\dot{\chi}}{\chi} = \beta - \alpha_o \chi - \frac{\gamma Z}{\chi} \tag{8}$$

where $\dot{\chi} = \frac{\partial \chi}{\partial t}$, $\beta = \ell n(AP)$, and α_o, γ are renormalized coefficients.

The system of two interacting populations is described by the system of two equations. However, it is rather difficult to derive the equation which would describe changes in entomophage numbers. The present method allows one, by using only one equation for a phytophage (i.e. analyzing the behavior of the curves in the y, Θ plane by means of the difference $y_\beta - y_\alpha$ and the derivative $\frac{\partial y}{\partial \Theta}$), to reveal the ability of a given species to develop an outbreak under certain conditions. One will be able also to determine the relative contribution of inertial and noninertial regulating factors to the entire regulating system at a given stage of population development. This method will make it possible to estimate the efficiency of natural enemies and to find appropriate biological methods of control for a forest pest.

If entomophages do not cease to exert their regulatory effect on phytophages, whatever phytophage reproductive rates, and if the steepness in the $\theta' \leq \theta \leq \theta''$ portion of the $y=y(\theta)$ curve at $\chi=\chi_1$ is such that the inequality (5) is true, these entomophages will be promising biological agents against a forest pest. Such entomophages are, virtually, the major regulating factors which enable a scarce population to be kept at a stable level. However, when fluctuations in a pest population are large and the population has developed to outbreak proportions, these entomophages cease to exert their regulatory effects. In order that management policy directed towards the suppression of an outbreak be successful, it is necessary that the numbers of natural enemies be increased, i.e. food availability should be reduced to the threshold value θ'.

Thus the present method of analysis of $y=y(\chi,\theta)$ in the y,θ phase plans permits one, with the help of an experiment, to determine such entomophage densities which would be useful for controlling a forest pest number.

A PHASE PORTRAIT OF A PHYTOPHAGE POPULATION

The $y=y(\chi,\theta)$ curves in the y,χ plane can be used to construct a phase portrait of a population in the y,χ plane, which would be typical for a population in a given biogeocenosis.

It is a well established fact that in a forest biogeocenosis reproductive rates and population density are limited to certain values. So the problem is to find boundary values of the phase portrait in this plan. When fluctuations in modifying factors are small, the boundaries for the phase portrait will be represented by the curves which are built by typical points of the y,θ phase plane. If one plots in the y,χ phase plane points of intersections of the $y=y(\theta)$ curves with the axis y (in the y,θ plane) those points will indicate the lower boundary in the phase portrait (the curve $y=y'$) (Figure 3). Intersects of horizontal asymptotes with the y axis at different χ values will indicate the upper boundary in the phase portrait (the curve $y=y''$) (Figure 3). The biological significance of these curves is that they bound the phase area within which the system "phytophage-biogeocenosis" can function, provided that fluctuations in modifying factors are small. The system can leave these boundaries for a period of time which is much shorter than the generation time of a pest, as noninertial factors regulating the system beyond these boundaries force the system to return to the boundaries of the phase portrait.

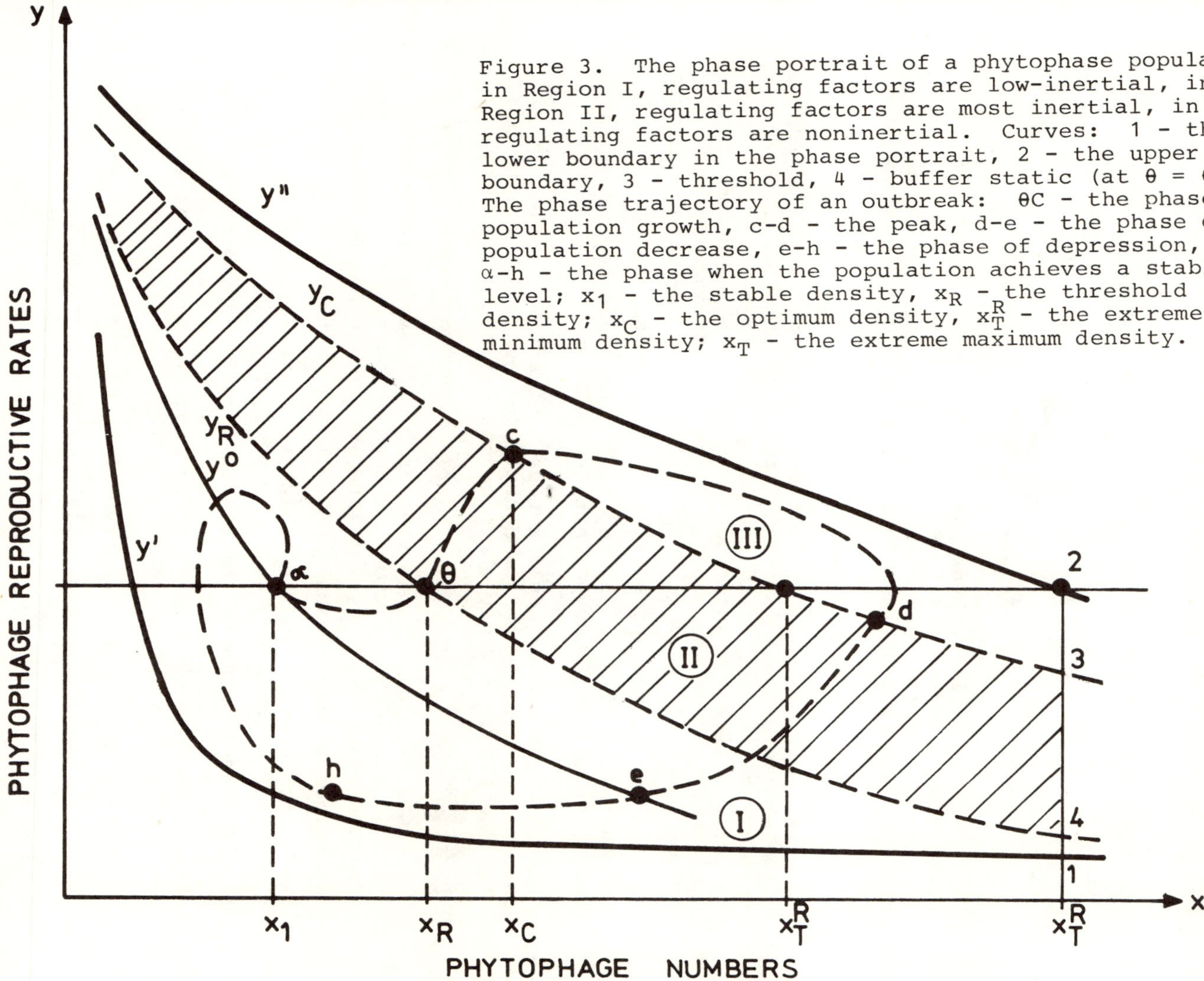

Figure 3. The phase portrait of a phytophase population in Region I, regulating factors are low-inertial, in Region II, regulating factors are most inertial, in III, regulating factors are noninertial. Curves: 1 - the lower boundary in the phase portrait, 2 - the upper boundary, 3 - threshold, 4 - buffer static (at $\theta = \theta'$). The phase trajectory of an outbreak: θC - the phase of population growth, c-d - the peak, d-e - the phase of population decrease, e-h - the phase of depression, α-h - the phase when the population achieves a stable level; x_1 - the stable density, x_R - the threshold density; x_C - the optimum density, x_T^R - the extreme minimum density; x_T - the extreme maximum density.

These boundary values have one important peculiarity: a set of points that form the lower boundary in the phase portrait defines a maximum, negative lag of inertial factors while the points in the upper boundary defines a maximum positive lag. A static curve, $y=y_o$, whose lag is zero, divides the phase portrait into two sectors of a positive and negative lag, respectively. The equilibrium point (χ_1) of the system under study results from the intersection of the $y=y_o$ curve with the straight line y=1.

The boundaries of the phase portrait are typical for a given biogeocenosis, provided that fluctuations of modifying factors are small. However, they can be drastically changed during the period of succession of an ecological system, and it is quite natural that each biogeocenosis has its own characteristic phase portrait.

If the relative contribution of the regulatory effect of entomophages becomes less and less important, the lower boundary curve would tend to the upper one. Hence, the phase volume would tend towards zero. Then the equilibrium point χ_1 would move across the straight line y=1 and achieve its boundary value χ i.e. $\chi_1 = \chi_T$ (Figure 3). The situation when χ_1 tends to χ_T is typical for a biogeocenosis in which, under the influence of man, entomophage numbers are decreased and, hence, their regulatory effect is weakened.

If entomophages could act as noninertial regulating factors, then both lower and upper curves would tend to the static curve y_o. Under these conditions the phase volume would tend to zero as well.

Thus, if one assumes that an ecological system has no inertial regulating factors, the phase portrait of this system would be represented by the line $y=y_o$. The phase volume is determined by the presence of inertial factors ($y_\beta - y_\alpha \neq 0$) while its location in the y,χ phase plane is determined by the relative contribution of these factors ($y_\beta - y_\alpha$). This enables one to measure the relative contribution of inertial factors to the entire regulating system, which is of great importance for exploration of appropriate biological methods of control of a forest pest number.

Let δ denote the relative width of the portrait at $\chi = \chi_1$, and y_α, y_β denote, respectively, maximum reproductive rates of a given population, then

$$\delta = \frac{2(y_\beta - y_\alpha)}{y_\beta - y_\alpha} \quad . \qquad (9)$$

Noninertial regulating factors are the main contributors to the entire regulation when $\delta \ll 1$. If $\delta \sim 1$ inertial factors become more important (natural enemies).

THE STRUCTURE OF THE PHASE PORTRAIT

The boundaries of the phase portrait are defined by the amount of contribution made by inertial factors to the entire regulating system, and its structure is defined by the nature of the interaction between insects and their natural enemies. As shown above, these interactions experience qualitative changes in the threshold points of food supply (θ', θ''), and these changes are characterized by the derivative $\frac{\partial y}{\partial \theta}$. A set of points θ' and θ'' in the y,θ plane is represented by two curves in the y,χ plane dividing the phase portrait into three sectors (Figure 3). The nature of the interaction between a species and its natural enemies is different in kind in each sector, and is dependent upon the position of the curve y_o in the y,χ plane.

The region in which regulating factors act as the most inertial ones (Region II) is bounded by curves 3,4 which correspond to a set of points θ', θ''. In this region insects can easily escape their natural enemies, as it is a unique region in which there emerges the positive feedback responsible for pest population growth.

In Region I the regulating effect is produced by biocenotic factors which are low-inertial since $\tau < T$.

Intraspecific noninertial factors act in Region III. Due to the absence of a lag in this region during an outbreak, or in the case of large fluctuations, reproductive rates of a population "are fading".

The structure of the phase portrait is defined by ecological peculiarities of a species and the nature of the interaction between this species and its natural enemies. The following situations are possible:

(I) If entomophage numbers in Region II are rigidly controlled by inertial regulating factors, i.e. the inequality (5) would never be true, then all the three regions in the phase portrait would be domains of stability, and the intersects of the y_o curve with the y=1 straight line would be the equilibrium point χ_1. Under these conditions the system drifts in the vicinity of the equilibrium point, the maximum speed of this drift taking place in Region II.

This structure of the phase portrait is typical to most forest insect phytophagous species, namely, to indifferent species of a relatively stable density, and species whose numbers can change over a wide range. It should be noted, however, that these species (e.g. most xylophagous species) are not capable of developing an outbreak in the full sense of the word, as changes in their population densities are determined by the amplitude of fluctuations in modifying factors. When fluctuations are completed, the regulating factors immediately return the system to a stable level. A typical feature of these systems is that positive feedback does not emerge when there is a rise in population density. That is why the population is unable to completely escape control by its natural enemies and develop an outbreak.

(II) If phytophage numbers are not rigidly controlled by inertial regulating factors, i.e. the inequality (5) is true, then in Regions II and III an outbreak may occur. In that case curve 3 would correspond to the threshold curve y_R, curve 4 - the buffer curve which restricts the development of an outbreak by maximum reproductive rates (Y_C), and curve 2 - the upper boundary of the phase portrait (Figure 3). In this case a domain of stability may occur only in Region I.

Such a phase portrait is typical of species which are able to escape control by natural enemies, and to increase significantly in numbers. These species are the most hazardous pests, since during an outbreak they may cause great damage to the forest biogeocenosis and even completely destroy it. An outbreak development in these species differs in kind from that developed in the above species. This difference is associated with peculiarities of the interaction between the population and the forest ecosystem.

THE TRAJECTORY OF AN OUTBREAK IN THE PHASE PORTRAIT

Consider the regularities of movement of a population across the phase trajectory of an outbreak, taking into consideration the effects of inertial and noninertial regulating factors. Our previous work has demonstrated that in an undisturbed, stable biogeocenosis, insect populations (including species capable of developing outbreaks) are scarce, i.e. their numbers are kept at a stable, low level (Isaev, Khlebopros, 1973, 1974). Reproductive rates of these scarce populations are close to unity (Figure 3). When external fluctuations are absent, the system can exist at point X_1 for an indefinitely long period of time, since the rate of change in numbers, as well as acceleration of numbers, will be equal to zero. If external fluctuations shift a population out of the equilibrium, the system sooner or later would return to X_1 under pressure of the negative feedback, as the period of time lag of regulating factors in this instance is the shortest.

If the amplitude of fluctuations in modifying factors (weather conditions or food supply) is large enough to throw a population out of the domain of stability to above the threshold curve y_R, an outbreak is then inevitable. During an outbreak, the population goes through five different stages: the phase of population growth; the peak phase; the phase of population decrease; the phase of depression; and the phase of population increase, when the population is stabilized at a certain level (Isaev, Khlebopros, 1973).

As seen in Figure 3, the phase of population growth (the Θ-C portion) corresponds to Region II in which regulating factors are highly inertial. In this region, under outbreak conditions, the negative feedback changes for the positive one, and the period of time lag associated with the effects of regulating factors is the greatest. This enables a population to escape the control of its natural enemies; population density as well as reproductive rates will be rapidly increasing. If factors that restrict the movement of the system were absent, insect numbers would increase infinitely because of the inability of regulating factors to reassert control over the population again. In Region III the growth of the population is limited by the effect of noninertial factors. Point C corresponds to maximum reproductive rates. This is the population optimum of an outbreak, associated with qualitative changes in the population state.

It is to be noted that in the phase of population growth, positive feedback emerges not only because the population is able to escape its natural enemies, but also because there are intraspecific factors which bring about an increase in pest numbers. The detailed analysis of Dendrolimus sibirica T outbreaks showed that at the phase of population growth in the original centers of infestation, the population density is increasing; at the same time, pest populations are concentrated at the expense of immigration from neighboring areas (Kondakov, 1974). The dynamics of population density is drastically changed when a part of the population takes an accelerated one-year cycle. This results in a 100-fold increase in the pest population density hundreds of times.

The C-d portion (Figure 3) should be considered as the peak phase of an outbreak. This phase is realized in Region III in which prevail noninertial, regulating factors. Under the pressure of these factors there takes place a change in a sign of the feedback. Reproductive rates decrease, and the period of time lag becomes shorter. Therefore, natural enemies begin "to catch up" with the prey, whose growth is slowed down by the effect of noninertial regulating factors. The effect of noninertial factors result in an important decrease in pest reproductive rates. For example, at the peak phase strong competition reduces fecundity and survival of *Dendrolimus sibirica* T. by 1.5 and 2.5 times, respectively.

Intraspecific regulating factors begin to operate so as to decrease pest population density. At this stage of an outbreak, the *Dendrolimus sibirica* population begins to migrate from the centers of infestation. The sex ratio is changed in favor of males (56 percent). The densities are then controlled mainly by specialized parasites.

If one assumes that in Region III entomophages and diseases are not able to control population density, ($\frac{X}{Z} \to \infty$), then the trajectory of an outbreak at the peak phase would come to its close at point X_T, at which the population density is stabilized at the level determined by the effect of noninertial factors only. If natural enemies produce their effect on the population as quickly as possible, the phase trajectory in the final part of the peak phase will cross the threshold point X_T^R. Each particular case is completed between these two points after crossing the buffer curve (point d, Figure 3).

At the phase of population decrease (de portion), inertial factors play a special role. Actually, they take up the torch from noninertial factors, the effect of which failed to return the system into the initial state. Due to the presence of natural enemies, which increase in numbers faster than the host-phytophage, the trajectory of an outbreak intersects the straight line y=1 and, at the point of intersection, the rate of pest population growth is equal to zero, and the negative feedback has a maximum value.

The phase of population decrease corresponds to Region II, in which the effect of inertial factors becomes progressively more important. This is associated with a sharp increase in the population density of natural enemies and a reduced amount of food available for them. Consequently, at the phase of population decrease, phytophage numbers are progressively decreasing. The effect of intrapopulation factors at this stage also results in a decrease in population density. *Dendrolimus sibirica* fecundity in this phase is reduced 1.3 fold, the sex ratio is changed in favor of males and the percentage of individuals taking a two-year developmental cycle increases. Owing to increased pressure of entomophages and massive epyzooty, survival is decreased 5 or 10 times (Kondakov, 1975).

Movement of the system in Region II, at both phases of population growth and decrease, is characterized by one common property, namely, by the presence of positive feedback. At the phase of population growth, the positive feedback makes it possible for the population to escape its natural enemies, while at the phase of population decrease it allows natural enemies to take up control over the population again. It is significant that the rate of population growth can be compared with the rate at which the population becomes scarce (Kondakov, 1954).

After crossing the threshold curve y_R the population returns to the domain of stability (Region I). However, the population goes through the phase of deep depression (e-h portion) before its density achieves a stable level. This phase is characterized by very low reproductive rates and a maximum pressure of entomophages. This high pressure of entomophages is associated with the positive feedback, which is still present, and with a rapid decrease in the period of time lag. For example, due to extremely unfavorable conditions, Dendrolimus sibirica during the phase of depression has very low survival (less than 1 percent), and the lowest reproductive rates (y=0,01-0,05). After this phase, the population begins to gradually return to a stable level. Then reproductive rates are rapidly increasing (in Dendrolimus sibirica they increase 100 fold in a two-year period) and after a while, population density achieves a stable level. This final phase of the outbreak can be considered as the phase at which a species stabilizes its density (h-a portion).

It is clear from the above that an outbreak may occur in a system in which the population has a domain of stability and is capable of maintaining its numbers at point χ_1. Such a phase portrait will be characteristic of most forest insect species (mainly Lepidoptera sp.) which are able to develop outbreaks. In Siberian forests, this kind of outbreak is characteristic of Ocneria dispar L., Dendrolimus superans sibiricus T., Bupalus piniarius L., Boarmia bistortata Goere, Orgyia antiqua I. and a number of other pest species. These species may develop to outbreak proportions only when fluctuations in modifying factors are large enough to push the system out of the threshold curve y_R into the region in which regulating factors are most inertial. As a rule, such outbreaks spread and contaminate progressively larger areas, their emergence being brought about by cyclic changes in weather conditions. After an outbreak, populations of such species return to a stable level and for some time they keep their density stable.

However, it is quite possible in real world situations that an outbreak may be permanent in nature. In the phase portrait of such outbreaks, the static curve y_o is shifted from the domain of stability (Region I) into the region of an outbreak (Region II). If $\chi_1 \geq \chi_R$ then the intersect of the static curve y_o with the straight line y=1 should not be considered as the point of a stable density, or Region I as the domain of stability.

Under these conditions, a phytophage population may actually maintain an outbreak for a long time, i.e. the pest species does not stabilize its density at point χ, after the whole cycle of an outbreak is completed. Therefore, even when fluctuations in modifying factors are small, the population develops the next outbreak and goes through all the five phases in the phase trajectory.

In order that such systems could exist, the effect of these insects on a forest biogeocenosis should be relatively low (i.e. the insects must not push the forest ecosystem out of equilibrium). An outbreak of this kind can be demonstrated with the larch bud moth (*Zeiraphera diniana* Cin.) in the Swiss Alps. This pest causes great damage to the crown of larch during foliage development. However, it does not kill the tree since it grows needles for a second time. Baltensweiler demonstrated that larch bud moth outbreaks are permanent in nature, each lasting 8-10 years (Baltensweiler, 1970). Under favorable conditions for the bud moth (belt of forests 1700-1900 m altitude), there is a periodic occurrence of defoliation over an 8-year period. In the suboptimum area (750-900 m altitude), defoliation has no such distinct periodicity, since extreme bud moth population densities at the peak phase are relatively low. In the lower part of the forest belt, larch distribution becomes more patchy. In this instance, the larch bud moth population does not develop to outbreak proportions but remains at low stable densities. Similar observations were reported for the larch bud moth occurring in mountain forests in the south of Siberia. This insect also develops permanent outbreaks, each lasting 8-10 years (Florov, 1952).

The analysis of the phase trajectory of two full population cycles of an outbreak which occurred from 1949-1969 was described by Baltensweiler (1970). This analysis showed that the bud moth population, under the most favorable conditions, does not stabilize its density in the domain of stability. After the full cycle of an outbreak is completed, the influence of small fluctuations causes the next outbreak to develop. It is to be noted that this particular situation is likely to occur when the domain of stability in the phase portrait is very small. Then the population can rapidly develop an outbreak even under the influence of small fluctuations.

It is significant that maximum bud moth densities in cycles of a permanent outbreak are about the same in a particular habitat and similar ecotopes. This results from the fact that maximum bud moth densities under these conditions are limited by a shortage of food, due to complete defoliation of trees. At the phases of depression and stable population level, the bud moth population has different densities in different periods of time and in different biotopes.

In Baltensweiler's opinion, genetic polymorphism is an important factor regulating bud moth numbers. This intra-population, inertial factor responds phenotypically to increased bud moth numbers at the peak phase of an outbreak. The effect of such a factor is transformed through modifying factors.

Since dark ecotypes, which are most sensitive to qualitative composition of food, predominate in the population, the peak phase in the period of defoliation is completed more rapidly.

SUMMARY AND CONCLUSIONS

The analysis of forest insect population dynamics allows one to distinguish two types of regulating factors—inertial and noninertial. Inertial factors are functionally dependent on the population density of the previous generation and hence, they may have a long relaxation period. The effect of non-inertial (or low-inertial) factors is defined by the population density of the present generation. They have, actually, no period of time lag since they produce their regulatory effect during the period of one generation. The presence of different inertial factors makes it possible for a phytophage to escape its natural enemies and to develop an outbreak. The effect of noninertial factors allows reproductive rates of a population during an outbreak to be stabilized, and makes it possible for inertial factors to resume their regulatory effect. The qualitative nature of the population dynamics of different forest insect species is associated with the amount of contribution made by either of these factors.

The use of typical points of the system and individual estimation of regulatory factors, according to the degree of their inertial possibilities, have enabled the phase portrait of a population to be described, reflecting peculiarities of the ecology of a species in the system of a forest biogeocenosis. When constructing the phase portrait, we had to omit the effect on the population of important changes in modifying factors. This seemed a necessary condition, otherwise the population dynamics of the forest insect could not be captured. The analysis of the phase portrait structure enables one to establish qualitative and quantitative parameters for forest insect population dynamics, to measure the contribution of individual regulating factors, and to define conditions under which a species is able to develop an outbreak.

References

Florov, D.N. (1952), Larch Tortrix, *Izvestie RGO*, 84, vyp. 6 (in Russian).

Gershenzon, S.M. (1959), The Role of Ecological Factors in the Emergence of Viral Epizootic Among Lepidopteran Insects, *Voprosy ekologii*, 3, 20-24, Kiev (in Russian).

Hitchinson, G.E. (1948), Circular Causal System in Ecology, *Ann. N.Y. Acad. Sci.*, 50, 221-246.

Ianovskii, V.M., V.V. Kiselev (1975), The Role of Biotic Factors in the Regulaton of Larch Bark Beetle Populations, *Izvestiia SO AN SSSR, Seriia biolog*, 48-54 (in Russian).

Isaev, A.S., R.G. Khlebopros (1973), Principles of Stability in Wood Insect Population Dynamics, *Doklady AN SSSR*, Vol. 208, vyp. 1, 225-228 (in Russian).

Isaev, A.S., R.G. Khlebopros (1974), Analysis of Wood Insect Population Dynamics on the Basis of the Principle of Stability of Mobile Ecological Systems, *Zhurnal obshchei biologii*, Vol. 35, No. 5, 737-774 (in Russian).

Isaev, A.S., R.G. Khlebopros, Iu.P. Kondakov (1974), Regularities in Wood Insect Population Dynamics, *Lesovedenie*, No. 3, 27-42 (in Russian).

Iwao, S. (1962), Studies on the Phase Variation and Related Phenomena in Some Lepidopterous Insects, *Mem. College Agric.*, Kyoto Univ., N 84, 80.

Janisch, E. (1959), Populations Analysis Among Harmful Insects, Magazine of *Applied Entymology*, 43, H. 4, 371-386.

Kalomiets, N.G. (1962), *Parazity; Khishchniki sibirskogo shelkopriada* (Parasites and Predators of the Siberian Silkworm Moth), Izdatelštvo SO AN SSSR, Novosibirsk, 172.

Kondakov, Iu.P. (1974), Regularities in the Mass Reproduction of the Siberian Silkworm Moth, in *Ekologiia populiatsii lesnykh zhivotnykh Sibiri* (The Ecology of Populations of Forest Animals of Siberia), Naulsa, Novosibirsk, 118-165 (in Russian).

Kostirin, V.A. (1937), Biology Mathematics, Paris.

Naumov, N.P. (1963), *Ekologiia zhivotnykh* (Animal Ecology), Vysshaia Shkola, Moscow, 618.

Nikitiuk, A.I. (1951), Predatory and Parasitic Insects as Regulators of the Pernecious Activity and Spread of Bark Beetles of Coniferous Forests *Sooboobshchenie I-Biull. MOIP otd. biol.*, 56, No. 5, 58-63 (in Russian).

Prozorov, S.S. (1940), The Series Larch Tortrix in the Eastern Saian in *Listvennitsa sibirskaia* (Siberian Larch), Krasnoiarsk (in Russian).

Rozhkov, A.S. (1965), *Massovoe razmnozhenie sibirskogo shelkopriada i mery bor'by s nim* (Mass Reproduction of the Siberian Silkworm Moth and Measures to Combat It), Nanka, Moscow, 178.

Schwerdtfeger, F. (1957), Forest Insects in Primeval and Commercial Forests, *General Forestry Journal*, 485.

Sharov, A.G., The Phase Character of Variability of Elm Collembola Larvae, *Doklady AN SSSR*, 88, 5, 929-932 (in Russian).

Solomon, M.E. (1949), The Natural Control of Animal Population, *Animal Ecology*, 18, N 1, 1-35.

Solomon, M.E. (1957), Dynamic of Insect Populations, *Annual Rev. Entomol.*, 2, 121-142.

Solomon, M.E. (1964), Analysis of Processes Involved in the Natural Control of Insects, *Advances Ecol. Res.*, 2, Acad. Press, London-N.Y., 1-58.

Steinhaus, E.A. (1958), Growing as a Possible Factor in Insect Disease, *Ecology*, 39, N 3, 503-514.

Takahashi, F. (1964), Reproduction Curve With Two Equilibrium Points: A Consideration on the Fluctuation of Insects Population, *Res. Population Ecol. V.*, 6, N 1.

Viktorov, G.A. (1965), Fluctuations in Insect Populations as a Regulating Process, *Zhurnal obshchei biologii*, Vol. 26, No. 1, 43-55 (in Russian).

Viktorov, G.A. (1967), *Problemy dinamiki chislennosti; nasekomykh na primere vrednoi cherepashki* (On the Dynamics of Insect Populations, Taking Eurygaster Intergriceps as an Example), Nauka, Moscow, 271.

Volterra, V. (1931), Lectures on the Mathematical Theory of the Struggle for Life, *Cahiers Scientist*, N. 7, Gauthiers-Villars, Paris, 214.

Wilbert, H. (1961), On the Determination and Control of the Average Density of Insect Populations, *Z. Morph. Okol. Tiere*, 50, H. 5, 510-519.

Zinov'ev, G.A. (1957), Materials for a Study of the Role of Biotic Factors in the Regulation of Populations of the Pest "Skrytostvol", *Entomologicheskii obzor*, 36, no. 2, 322-354 (in Russian).

Zwölfer, W. (1963), On the Means of Defense of our Forest Trees Against Insect Damage, *Z. Ang. Entomol.*, 51, H. 4, 364-370.

APPENDIX 1:

Participants and Authors of Conference Presentations

A. Bazykin
International Institute for
Applied Systems Analysis
Resources and Environment Area
Schloss Laxenburg
2361 Laxenburg, AUSTRIA

M. Birley
Imperial College Field Station
Silwood Park
Ascot, Berkshire SL5 7PY, UK

P.F. Boisvieux
Department of Biomathematics
C. H. U. Pitie - Salpetriere
89 avenue de la Gare
75013 Paris, FRANCE

H. van den Bos
Institut für Forstzoologie
Universität Göttingen
Buesgenweg
3400 Göttingen-Weende, FRG

J. Celecia
Division of Ecological Sciences and MAB
United Nations Educational, Scientific
and Cultural Organization (UNESCO)
Place de Fontenoy
Paris 7, FRANCE

W.C. Clark
Institute of Resource Ecology (IRE)
University of British Columbia
Vancouver, 8, British Columbia, CANADA

H.N. Comins
Imperial College Field Station
Silwood Park
Ascot, Berkshire SL5 7PY, UK

G.R. Conway
Imperial College Field Station
Silwood Park
Ascot, Berkshire SL5 7PY, UK

G.L. Curry
Texas A&M University
Department of Industrial Engineering
College Station, Texas 77801, USA

D.W. DeMichele
Texas A&M University
Biosystems Research Division
Department of Industrial Engineering
College Station, Texas 77801, USA

E. Donaubauer
Institute of Forest Protection
Tirolergarten
1131 Vienna, AUSTRIA

A.P. Gutierrez
University of California, Davis
Department of Entomology
Davis, California 95616, USA

Z. Harnos
Computing Center
Hungarian National Planning Office
Victor Hugo u. 18-22
Budapest 13, HUNGARY

J.C. Headley
University of Missouri
Department of Agricultural Economics
Columbia, Missouri 65201, USA

I. Hodek
Institute of Entomology
Czechoslovak Academy of Sciences
Vinicna 7
Prague 2, CZECHOSLOVAKIA

C.S. Holling
Institute of Resource Ecology (IRE)
University of British Columbia
Vancouver, 8, British Columbia, CANADA

A.S. Isaev
c/o The Academy of Sciences
State Committee for the USSR
Council of Ministers for
Science and Technology
11 Gorky Street
Moscow, K-9, USSR

D.D. Jones
Institute of Resource Ecology (IRE)
University of British Columbia
Vancouver, 8, British Columbia, CANADA

R.G. Khlebopros
c/o The Academy of Sciences
State Committee for the USSR
Council of Ministers for
Science and Technology
11 Gorky Street
Moscow, K-9, USSR

K. Kiritani
Kochi Prefectural Institute
of Agricultural and
Forest Science,
Ino, Agawa-gun,
Kochi 781-21, JAPAN

N.Y. Konovalov
c/o The Academy of Sciences
State Committee for the USSR
Council of Ministers for
Science and Technology
11 Gorky Street
Moscow, K-9, USSR

L. Melnikov
c/o The Academy of Sciences
State Committee for the USSR
Council of Ministers for
Science and Technology
11 Gorky Street
Moscow, K-9, USSR

C. Millier
National Center for Forest Research (CNRF)
National Institute for Agronomic Research (INRA)
Sation de Biometrie
54280 Champenoux, FRANCE

G.A. Norton
Imperial College Field Station
Silwood Park
Ascot, Berkshire SL5 7PY, UK

V.J. Petrinsky
Federal Ministry of Health and
Environmental Protection
Stubenring 1
1010 Vienna, AUSTRIA

G. Pühringer
Institut für Experimentelle Zoologie
Universität für Bodenkultur
Gregor Mendel Strasse 33
1180 Vienna, AUSTRIA

R. Rainey
Centre for Overseas Pest Research
Ministry of Overseas Development
Wrights Lane
London W8, UK

U. Regev
Ben-Gurion University
Department of Economics
Beersheva, ISRAEL

C. Sanders
Canadian Forest Service
Great Lakes Forest Research Center
P.O. Box 490
Sault Ste. Marie, Ontario, CANADA

H. Schieferdecker
c/o Academy of Sciences of the GDR
Institute of Geography and Geo-Ecology
Rudower Chaussee 5
1199 Berlin-Adlershof, GDR

H. Shalit
University of California, Berkeley
Department of Agricultural and
Resource Economics
Berkeley, California 94720, USA

C.A. Shoemaker
Cornell University
Department of Environmental Engineering
Ithaca, New York 14853, USA

Y. Wang
University of California, Berkeley
Division of Biological Control
1050 San Pablo Avenue
Albany, California 94706, USA

H.-H. Wilfert
c/o Academy of Sciences of the GDR
Central Research Institute for Cybernetics
and Information Processes
Department of Technical Cybernetics
Haeckerstrasse 20
8027 Dresden, GDR

APPENDIX 2:

Pest Management Network

Dr. Altenkirch
Institut für Pflanzenpathologie
und Pflanzenschutz
Nikolausberger Weg 144
3400 Göttingen
FRG

R. Alves
Secretaria de Agricultura
Instituto Biológico
1252-Caixa Postal
7119 Sao Paulo
BRAZIL

D.M. Auslander
University of California,
Berkeley
Department of Mechanical
Engineering
1050 San Pablo Avenue
Albany, California 94706
USA

W. Baltensweiler
Eidgenössische Technische
Hochschule Zürich
Rämistrasse 101
8092 Zürich
SWITZERLAND

G.L. Baskerville
Maritimes Forest Research Centre
P.O. Box 4000
Fredericton, New Brunswick
CANADA

A. Berryman
Washington State University
Department of Entomology
Pullman, Washington 99163
USA

H.K. Bichtler
The Academy of Sciences of
the German Democratic Republic
Otto-Nuschke-Str. 22-23
Berlin
GDR

J.R. Blais
Laurentian Forest Research
Centre
P.O. Box 3800
Ste. Foy, Quebec GlV 4C7
CANADA

G. Briolini
Istituto di Entomologia
Universita degli Studi di
Bologna
Via Filippo Re, 6
40126 Bologna
ITALY

D. Burnell
Washington State University
Department of Entomology
Pullman, Washington 99163
USA

G. Carlson
North Carolina State University
Department of Economics
Raleigh, North Carolina 27607
USA

J. Carrizosa
Instituto de Desarrollo de los
Recursos Naturales (INDERENA)
Carrera 14-No 25A-666
Bogotá
COLOMBIA

D. Chant
University of Toronto
Toronto 181, Ontario M5S 1A1
CANADA

E. Clive
Rothamsted Experimental Station
Harpenden, Hertfordshire AL5 2JQ
UK

G. Cussans
The Weed Research Organization
Begbroke Hill
Sandy Lane
Yarnton, Oxford OX5 1PF
UK

C.T. de Wit
Agricultural University
Department of Theoretical
Production Ecology
De Dreyen 4
Wageningen
THE NETHERLANDS

V. Eidman
University of Minnesota
Department of Agricultural and
Applied Economics
St. Paul, Minnesota 55108
USA

W. Faber
Bundesanstalt für Pflanzenschutz
Trunnerstrasse 5
1020 Vienna
AUSTRIA

P. Fischer-Colbrie
Bundesanstalt für Pflanzenschutz
Trunnerstrasse 5
1020 Vienna
AUSTRIA

J. Franz
Institut für Schädlingsbekämpfung
Heinrichstr. 243
6100 Darmstadt
FRG

W.H. Fuchs
Institut für Pflanzenpathologie
und Pflanzenschutz
Nikolausberger Weg 14
3400 Göttingen
FRG

E. Haukioja
University of Turku
Department of Zoology
SF-20500
Turku 50
FINLAND

D.L. Haynes
Michigan State University
Department of Entomology
East Lansing, Michigan 48823
USA

C. Huffaker
International Center for
Biological Control
University of California,
Berkeley
1050 San Pablo Avenue
Albany, California 94706
USA

T. Huflejt
Warsaw Agriculture University
Department of Forest and
Wood Protection
Warsaw
POLAND

C. Ignoffo
Biological Control Laboratory
U.S. Department of Agriculture
Columbia, Missouri 65201
USA

T. Inoue
Entomological Laboratory
College of Agriculture
Kyoto University
Kita-Sirakawa
Sakyo-ku, Kyoto 606
JAPAN

Y. Ito
Okinawa Agric. Expt. Sta.
Sakiyama 4-222
Naha, Okinawa 903
JAPAN

S. Iwao
Laboratory of Applied
Entomology and Nematology
Nagoya University
Chikasa, Nagoya
JAPAN

T. Jermy
Plant Protection Research
Institute
Novenyvedelmi Kutato Intezet
1525 Budapest, Pf. 102
HUNGARY

J. Karwowski
Warsaw Agriculture University
Department of Forest and
Wood Protection
Warsaw
POLAND

H. Klomp
Zoological Laboratory
Agricultural University
Wageningen
THE NETHERLANDS

J. Kranz
Tropeninstitut
Schottstr. 2-4
6300 Giessen
FRG

T. Lewis
Department of Entomology
Rothamsted Experimental Station
Harpenden, Hertfordshire AL5 2JQ
UK

B. Luck
University of California, Riverside
Department of Entomology
Riverside, California 92502
USA

R. Luti
Centro de Ecologia y Recursos Naturales Renovables
Universidad Nacional de Córdoba
Avda. Vélez Sársfield 299
5000 Córdoba
ARGENTINA

R. May
Princeton University
Department of Biology
Princeton, New Jersey 08540
USA

J.M. McLeod
Laurentian Forest Research Centre
P.O. Box 490
Sault Ste. Marie, Ontario P6A 5M7
CANADA

K. Nakamura
Divison of Entomology
National Institute of Agricultural Sciences
2-1-7 Nishigahara, Kita-ku,
Tokyo 281
JAPAN

F. Nakasuji
Entomological Laboratory
College of Agriculture
Nagoya University
Furo-chyo, Chigusa-ku
Nagoya 468
JAPAN

M.M. Nielson
Maritimes Forest Research Centre
P.O. Box 4000
Fredericton, New Brunswick
CANADA

M.C. Nilsson
Lantbrukshogskolan/Vaxtskydd
Box 44
23053 Alnarp
SWEDEN

B. Ohnesorge
Sälzackerstr. 52
7000 Stuttgart-Möhringen
FRG

G.F. Oster
University of California, Berkeley
Department of Mechanical Engineering
1050 San Pablo Avenue
Albany, California 94706
USA

L. Pedigo
Iowa State University
Department of Entomology
Ames, Iowa 50010
USA

D. Pimentel
Cornell University
Department of Entomology
Comstock Hall
Ithaca, New York 14850
USA

H. Pschorn-Walcher
Commonwealth Institute of Biological Control
European Station
2800 Delemont
SWITZERLAND

R. Rabbinge
Agricultural University
Department of Theoretical Production Ecology
De Dreyen 4
Wageningen
THE NETHERLANDS

S.L. Readshaw
Commonwealth Scientific and Industrial Research Organization (CSIRO)
Division of Entomology
Canberra
AUSTRALIA

W. Rudd
Louisiana State University
Department of Computer Science
Baton Rouge, Louisiana 70803
USA

W. Ruesink
University of Illinois
Department of Entomology
163 Natural Resources Building
Urbana, Illinois 61801
USA

K. Russ
Bundesanstalt für Pflanzenschutz
Trunnerstrasse 5
1020 Vienna
AUSTRIA

H. Sandner
Zaklad ekologii PAN
Nowy Swiat 72
Warsaw
POLAND

T. Sasaba
Division of Entomology
Kochi Prefectural Institute of Agricultural and Forest Science
Ino, Agawa-gun
Kochi 781-21
JAPAN

J.W. Sinko
University of Rochester
Department of Electrical Engineering
Rochester, New York 14627
USA

R. Stinner
North Carolina State University
Raleigh, North Carolina 27607
USA

W. Streiffer
University of Rochester
Department of Electrical Engineering
Rochester, New York 14627
USA

R.W. Sutherst
Commonwealth Scientific and Industrial Research Organization (CSIRO)
Division of Entomology
Canberra
AUSTRALIA

Y. Takahashi
University of California, Berkeley
Department of Mechanical Engineering
1050 San Pablo Avenue
Albany, California 94706
USA

C.R. Taylor
Texas A&M University
Department of Agricultural Economics
College Station, Texas 77843
USA

R.L. Tummala
Michigan State University
Department of Electrical Engineering
East Lansing, Michigan 48824
USA

E.E. Turtle
Plant Protection Service
Plant Production and Protection Division
Food and Agriculture Organization of the United Nations (FAO)
Via delle Terme di Caracalla 00100
Rome
ITALY

Y. Wang
University of California,
Berkeley
Division of Biological Control
1050 San Pablo Avenue
Albany, California 94706
USA

M.J. Way
Imperial College Field Station
Silwood Park
Ascot, Berkshire SL5 7PY
UK

L. Weismann
Ústav exper. biológie a
ekológie SAV
ul, Obrancov mieru 1
88534 Bratislava
CZECHOSLOVAKIA

H. Wilbert
Stauffenbergring 23
3400 Göttingen
FRG

J.C. Zadoks
Agricultural University
Department of Phytopathology
Binnenhaven
Wageningen
THE NETHERLANDS

W. Zakrzewski
Institute for Natural Foundations
of Forestry and Forest Culture
Warsaw Agriculture University
Warsaw
POLAND

J. Zeddies
Universität Hohenheim
Angewandte Landwirtschaftliche
Betriebslehre
Stuttgart 70 (Hohenheim)
FRG

B. Zwatz
Bundesanstalt für Pflanzenschutz
Trunnerstrasse 5
1020 Vienna
AUSTRIA